The marine zooplankton is one of the most fascinating and diverse assemblages of animals known to biologists. This is a new edition of the successful students' manual providing a photographic guide to representative forms of the major groups from medusae and comb jellies to larval fish and squid. In it, only live and active organisms have been photographed, giving a unique visual perspective. In this new edition, the taxonomy and morphology have been revised and brought up to date, so that, combined with information on behaviour and development, this book creates a vivid and essential reference text for all those interested in marine zooplankton.

COASTAL MARINE ZOOPLANKTON
A practical manual for students

Nephrops norvegicus (L.). Stage 3 *zoea* larva. Length 10 mm.

COASTAL MARINE ZOOPLANKTON
A practical manual for students

C.D. Todd*, the late M.S. Laverack & G.A. Boxshall[†]
*Gatty Marine Laboratory
University of St Andrews
Scotland

[†]The Natural History Museum
Cromwell Road, London
England

Second edition

CAMBRIDGE
UNIVERSITY PRESS

CAMBRIDGE UNIVERSITY PRESS
Cambridge, New York, Melbourne, Madrid, Cape Town, Singapore, São Paulo, Delhi

Cambridge University Press
The Edinburgh Building, Cambridge CB2 8RU, UK

Published in the United States of America by Cambridge University Press, New York

www.cambridge.org
Information on this title: www.cambridge.org/9780521555333

First published 1991
Second edition 1996
Reprinted 2000, 2003, 2006

A catalogue record for this publication is available from the British Library

Library of Congress Cataloguing in Publication data
Todd, C. D.
Coastal marine zooplankton: a practical manual for students /
C.D. Todd, M.S. Laverack and G.A. Boxshall. – 2nd ed.
p.   cm.
Includes bibliographical references (p.   ) and index.
ISBN 0 521 55533 7 (pb)
1. Marine zooplankton – Classification.   2. Marine zooplankton – Identification.
I. Laverack, M. S.   II. Boxshall, Geoffrey Allen.
III. Title.
QL123.T64 1996
592.092–dc20   95-24608 CIP

ISBN 978-0-521-55533-3 paperback

Transferred to digital printing 2009

CONTENTS AND INDEX OF PHYLA

INTRODUCTION

This teaching manual has been produced with the aid of a generous grant from the Nuffield Foundation. It is hoped that the manual should satisfy a number of purposes and objectives, but there are certain functions which it is most certainly not intended to serve. Foremost among the latter is our express desire that students should not content themselves in identifying an organism by merely fitting the animal to a picture. There is no substitute for the use of the appropriate dichotomous keys and carefully illustrated and annotated taxonomic texts. Indeed, we have produced a manual that incorporates only examples or representatives of the major coastal zooplankton taxa. In so doing, we have attempted to provide illustrations of all the major groups likely to be encountered in typical plankton samples, as well as pictures of some which are rarely seen.

The timing of the traditional university academic year is such that many seasonal zooplankters will never be seen alive by students, even if provided with freshly caught samples. This unavoidable problem is, we hope, at least partially offset by some of our photographic material.

Our prime objective has been to photograph live plankters in their natural state (or as close to it as possible). Preserved zooplankters lose much of their colour, transparency and perhaps overall form or appearance. For example, gastropod molluscan veligers are never preserved with the velum and cilia extended: they appear simply as tiny coiled shells. Our photographs therefore go some way towards illustrating the living organism as it would be seen in a fresh sample. In seeking that objective we must, however, acknowledge that despite our efforts many of the pictures leave a great deal to be desired: these small, rapidly-moving, transparent organisms are not the most co-operative of subjects.

Students should also be aware that many organisms are, to a greater or lesser degree, damaged during the process of capture and sample sorting. Obviously, we have generally selected intact specimens for the manual, but you should remember that tentacles, spines, antennae, limbs etc. are frequently broken or absent in routine sample material. Indeed, it can be a rather challenging exercise to attempt to identify an organism from only a detached or damaged fragment; the dissociated comb plates of ctenophores are often variously ascribed to a whole host of taxa, ranging up to the Urochordata!

Throughout the manual, wherever possible, we have identified the organism and provided the appropriate taxonomic summary. Whatever one's ultimate purpose in studying zooplankton, it is imperative not to lose sight of taxonomy and basic zoology. It is most important to understand the systematics and taxonomy of these organisms because such knowledge permits evaluation of their biology, and thence perhaps their ecology. It is almost valueless simply to put a name-tag to an animal. However, in order to undertake quantitative exercises it is essential to be able to identify an organism to species level. Be warned: all copepods look the same at first, but careful observation will prove otherwise. We suggest that the manual is perused before your first sample is inspected. This may encourage you to believe that identification of the organisms is often not such a formidable task as may at first appear.

In general we have attempted to follow a phyletic sequence in the presentation of the photographs but restrictions on the number of pages, and a requirement to retain a uniform black or white background to all the photographs in any one figure, has necessarily led to some unfortunate mixtures. Wherever relevant, however, we have attempted to reduce this complication by appropriate cross-references to the figures. Those which are most out of place are the pictures of the dinoflagellate *Noctiluca*, which is hidden amongst annelids, and of the only fish we have included, which are found in Fig. 38 alongside *Phoronis*. We do not believe this affects the value of the presentation.

We are pleased to acknowledge the assistance of Sonny Darrig and Steve Hall in collating some of the original photographs. Staff at the Marine Biological Association, Plymouth, kindly assisted in the collection of some of the material photographed. A number of colleagues offered advice, corrections and additional suggestions for this revised and corrected edition. In this regard we are especially pleased to acknowledge the contributions of P.F. Clark, J.D. George, J. Martin and G. Walker.

Identification, taxonomy and suggested source material

Taxonomic descriptions of the myriad organisms of the zooplankton are contained in numerous disparate sources and no simple list can embrace all possible organisms which may be encountered. This is true even of small biogeographic regions such as the British Isles. There are, however, some general publications which provide a wealth of information on a wide variety of organisms. Foremost among these is Thorson's (1946) monograph. Many, but by no means all, zooplanktonic taxa of the North Atlantic have been covered in the continuous publication (since 1947) of the series of hand-sheets (*Fiches d'identification du Zooplankton*) under the aegis of the *Conseil international pour l'exploration de la mer*. These fiches are issued at irregular intervals with each covering a small portion of a taxonomic group. Sheet 1 deals with the entire Phylum Chaetognatha, but, more typically, sheet 2 concerns the hydromedusae of only one family (Tubulariidae) of hydroids. Such groups as the medusae, Copepoda, Decapoda, Polychaeta, larval forms etc. clearly require more than one sheet. The series remains incomplete and continues to appear. For those groups covered it is an invaluable reference guide. Fiches (and other texts) concerning the major planktonic organisms include:

Cnidarian medusae	**Russell, F.S.** (1953, 1970). The medusae of the British Isles. 2 vols. Cambridge University Press. (dates various) Fiche Nos. 2, 28–31, 51, 54, 99, 102, 128, 152–154, 158, 161, 164–166.
Siphonophora	**Totton, A.K.** (1965). A synopsis of the Siphonophora. B.M.(N.H.), London.
	Kirkpatrick, K.A. & P.R. Pugh (1984). Siphonophores and vellelids. Synopses of the British Fauna, No. 29. Linn. Soc. (E.J. Brill).
Ctenophora	**Greve, W.** (1975). Fiche No. 146.
Polychaeta	**Muus, B.J.** (1953). Fiche Nos. 52 and 53.
	Hamond, R. (1967). Fiche No. 113.
	Hannerz, L. (1961). Fiche No. 91.
Chaetognatha	**Fraser, J.H.** (1957). Fiche No. 1.
	Pierrot-Bults A.C. & Chidgey K.C. (1988). Synopses of the British Fauna, No. 39. Linn. Soc. (E.J. Brill).
Copepoda	**Sars, G.O.** (1903–1918). Crustacea of Norway, Vols. 4–8. Bergen Museum.
	Farran, G.P. (1948). Fiche Nos. 11–16, 32–40.
	Ogilvie, H.S. (1953). Fiche No. 50.
	Lovegrove, T. (1956). Fiche No. 63.
	Wells, J.B.J. (1970). Fiche No. 133.
	Corral, J. (1972). Fiche No. 138.
	Isaac, M.J. (1975). Fiche Nos. 144 and 145.
	Huys, R. & G.A. Boxshall (1991). Copepod evolution. Ray Society, London.
Mysidacea	**Nouvel, H.** (1950). Fiche Nos. 18–27.
	Makins, P. (1977) A guide to British Coastal Mysidacea *Field Studies*, 4, 575–595.
Amphipoda	**Dunbar, M.J.** (1963). Fiche No. 103.
	Shih C.-T. & M.J. Dunbar (1963). Fiche No. 104
	Lincoln, R.J. (1979). British Marine Amphipoda: Gammaridea. B.M.(N.H.), London.

Cirripedia (larvae)	**Lang, W.H.** (1980). Fiche No. 163.
Cumacea	**Jones, N.S.** (1957). Fiche Nos. 71–76.
	Jones, N.S. (1976). British Cumaceans. Synopses of the British Fauna, No. 7. Linn. Soc. (Academic Press).
Euphausiacea	**Mauchline, J.** (1971). Fiche Nos. 134–137.
	Mauchline, J. (1984). Euphausiid, Stomatopod and Leptostracan crustaceans. Synopses of the British Fauna, No. 30. Linn. Soc. (E.J. Brill).
	Baker, A. de C., Boden, B.P. & Brinton, E. (1990). A practical guide to the Euphausiids of the world. N.H.M., London.
Decapoda (larvae)	**Williamson, D.I.** (dates various). Fiche Nos. 67, 68, 90, 92, 109, 167, 168; (with **Pike, R.B.**), 81, 139; (with **Fincham, A.A.**), 159, 160.
	Ingle, R. (1991). Larval stages of northeastern Atlantic crabs: an illustrated key (Chapman & Hall).
Decapoda (adults)	**Rice, A.L.** (1967). Fiche No. 112.
	Smaldon, G. (1993). British coastal shrimps and prawns. Synopses of the British Fauna, No. 15. 2nd edition revised and enlarged by L.B. Holthuis & C.H.J.M. Fransen (Field Studies Council).
Phoronida (larvae)	**Forneris, L.** (1957) Fiche No. 69.
Bryozoa (larvae)	**Ryland, J.S.** (1965) Fiche No. 107.
Echinodermata (larvae)	**Mortensen, T.** (1931–1938). Contributions to the study of the development and larval forms of the echinoderms. 4 vols. Oxford University Press.
	Geiger, S.R. (1964). Fiche No. 105.
Hemichordata (larvae)	**Burdon-Jones, C.** (1957). Fiche No. 70.
Urochordata	**Fraser, J.H.** (1947). Fiche Nos. 9 and 10 (**Doliolids and salps**).
	Buckman, A. (1969). Fiche No. 7 (**Appendicularia**).
	Fraser, J.H. (1981). British pelagic tunicates. Synopses of the British Fauna, No. 20. Linn. Soc. (Cambridge University Press).
Larval monograph	**Thorson, G.** (1946). Reproduction and larval development of Danish marine bottom invertebrates, with special reference to the planktonic larvae in the Sound (Øresund). Meddelelser fra Kommissionen for Danmarks fiskeri- og Havundersøgelser, Serie Plankton, Vol. 4: 1–523.

In addition, there is an extremely useful (but rapidly dating) publication from the British Museum (Natural History) – **Sims, R.W.** (1980). *Animal Identification. A reference guide.* John Wiley & Sons – which provides primary sources for the identification of all groups in most biogeographic areas of the world.

A simple general account of Marine Zooplankton is given in the book of that name by **Wickstead , J.H.** (Institute of Biology Studies in Biology 62, Edward Arnold) published in 1976. A recently published book – **D.I. Williamson** (1992). *Larvae and evolution.* Chapman & Hall – outlines a highly controversial hypothesis relating to the evolution of certain planktonic larval forms (including specific crustacean zoea, hemichordate tornaria and echinoderm pluteus stages).

FIGURE 1: A–E Phylum Cnidaria, Class Hydrozoa

These five photographs illustrate a variety of cnidarian medusae. These are the pelagic phases of the life cycle of a range of otherwise benthic polypoid forms. In many cases medusae are given binomial scientific names, and treated as recognisable species in their own right, although for some the benthic polypoid form remains unknown. Invariably the only means of linking medusa to polyp is to rear the medusae released from known species of polyp in the laboratory. As is typical of most cnidarians, medusae are carnivorous and capture planktonic prey by means of nematocysts borne on the contractile tentacles.

Medusae should not be considered as larvae because they carry the gonads; they are differing structural forms within the life cycle, and alternate with the polypoid phase. The medusae are budded asexually by the benthic polyp and having completed growth and development in the plankton the medusae release haploid gametes into the water column. There, fertilisation gives rise to the ciliated planula larva, which settles to the substratum, attaches, and metamorphoses into the polypoid form. Growth of the polypoid form is by asexual budding of the colony. Medusae commonly are only a few millimetres in diameter, but may attain several centimetres across the bell. Most are transparent, gelatinous and fragile and swim by rhythmical contractions of the (longitudinal and circular) bell musculature. In preservative medusae lose their transparency and invariably become opaque.

FIG. 1A. *Aglantha digitale* (Müller). [length 10 mm]. This easily recognised medusa is one of the classic indicator species used in characterising water masses around the British Isles from zooplankton samples. The eight gonads (g) are situated on the radial canals at the junction of the peduncle (p) and subumbrella. st, stomach; m, mouth; mt, marginal tentacles.

FIG. 1B. *Amphinema rugosum* (Mayer). [length of bell 5 mm]. In contrast to *Aglantha*, *Amphinema* lacks a peduncle. Small marginal tentaculae (smt) are present, along with two large tentacles, which in *A. rugosum* are golden brown in colour. The umbrellar opening bears a thin membranous velum (v), which constricts the opening of the bell. The radial canals (ra.c) and ring canal (rc) are clearly visible in the photograph.

FIG. 1C. *Leuckartiara* sp. [length 3 mm]. This probably is a juvenile of the form illustrated in Fig. 2E: see also Fig. 3A,B.

FIG. 1D. *Neoturris pileata* (Forskål). [length of bell 15 mm]. The polypoid form of this medusa is unknown. Characteristic of *Neoturris* is the large stomach (lacking a peduncle) and elaborately crenulate mouth (m). Note the thickened and compressed appearance of the bases of the (contracted) marginal tentacles (mt).

FIG. 1E. *Amphinema dinema* (Péron & Lesueur). [length of bell 3 mm]. The stomach (st) of this striking medusa is bright green and the marginal tentacles (mt) are violet. The mouth (m) typically comprises four crenulate and recurved lips, which are clearly seen in the illustration.

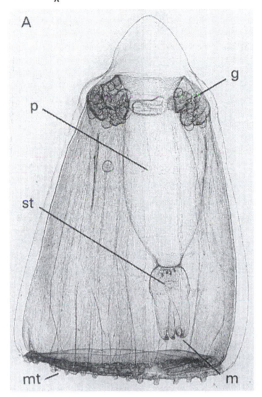

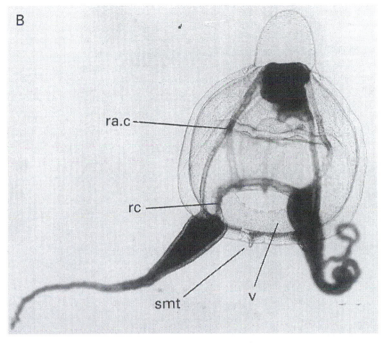

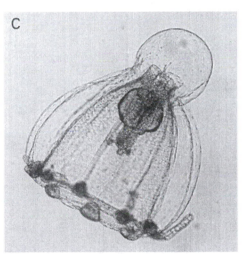

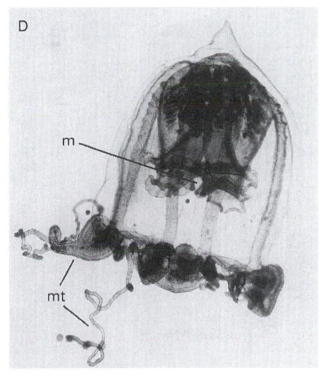

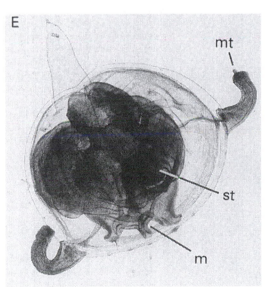

FIGURE 2: A–E Phylum Cnidaria, Class Hydrozoa

These five photographs illustrate other species of medusae. Many texts refer to anthomedusae – which possess gonads on the stomach and, for a short distance, on the radial canals and bear no statocysts – and leptomedusae – in which the gonads are on the radial canals, though they may reach the stomach, and statocysts are found ectodermally on the velum. Anthomedusae are invariably from athecate hydroids whilst leptomedusae are normally from thecate hydroids (rarely from naked hydranths).

FIG. 2A. *Eutonina indicans* (Romanes). [diameter 30 mm]. In the British Isles this species may be restricted to east Scotland and north-east England. Note the large number of small, equally sized marginal tentacles. This particular photograph is of a subumbrellar view and the thickness of the umbrellar jelly is not apparent. In fact the jelly thickness is rather similar to that of *Eutima gracilis* (Forbes & Goodsir) (Fig. 2B). The extensive peduncle, bearing the small stomach (st) and elaborate mouth can be clearly seen, as can the four gonads (g), which typically are located on the subumbrellar surface beneath the radial canals.

FIG. 2B. *Eutima gracilis* (Forbes & Goodsir). [diameter 10 mm]. This medusa is characterised by the pair of large tentacles and the gonads restricted to the peduncle (p) – they do not extend to the subumbrella. The typical hemispherical shape can be seen in the upper specimen and the four thin radial canals in the lower specimen. In both specimens the numerous small marginal swellings (ms) on the umbrella periphery can also be seen. The jelly of the umbrella is comparatively thick and solid.

FIG. 2C. *Bougainvillea* sp. [diameter 3 mm].

There are several species of *Bougainvillea*, which are generally separable according to the form of the tentacles and whether or not there is a peduncle present. In the illustrated specimens the thickness of the umbrella jelly and shortness of the marginal tentacles is apparent. Also clearly visible in the right-hand individual are the four subumbrellar gonads.

FIG. 2D. *Podocoryne* sp. [diameter 1 mm]. This small medusa is probably *Podocoryne*. The overall appearance is not dissimilar to *Bougainvillea*, but the differences in the size, number and distribution of the marginal tentacles are obvious. The bases, or bulbs (mtb), of the marginal tentacles are a bright red-brown in coloration.

FIG. 2E. *Leuckartiara* sp. [height of bell 7 mm]. This specimen is possibly a young *Leuckartiara octona* (Fleming) (see Fig. 3A,B). The genus is characterised, amongst other features, by an umbrella bearing an apical process of variable shape, the absence of a peduncle, and a crenulated mouth. The various species of *Leuckartiara* are separable (in the fully developed form only) by the number of marginal tentacles. Note the extensive velum (v) which constricts the opening of the bell in the individual illustrated.

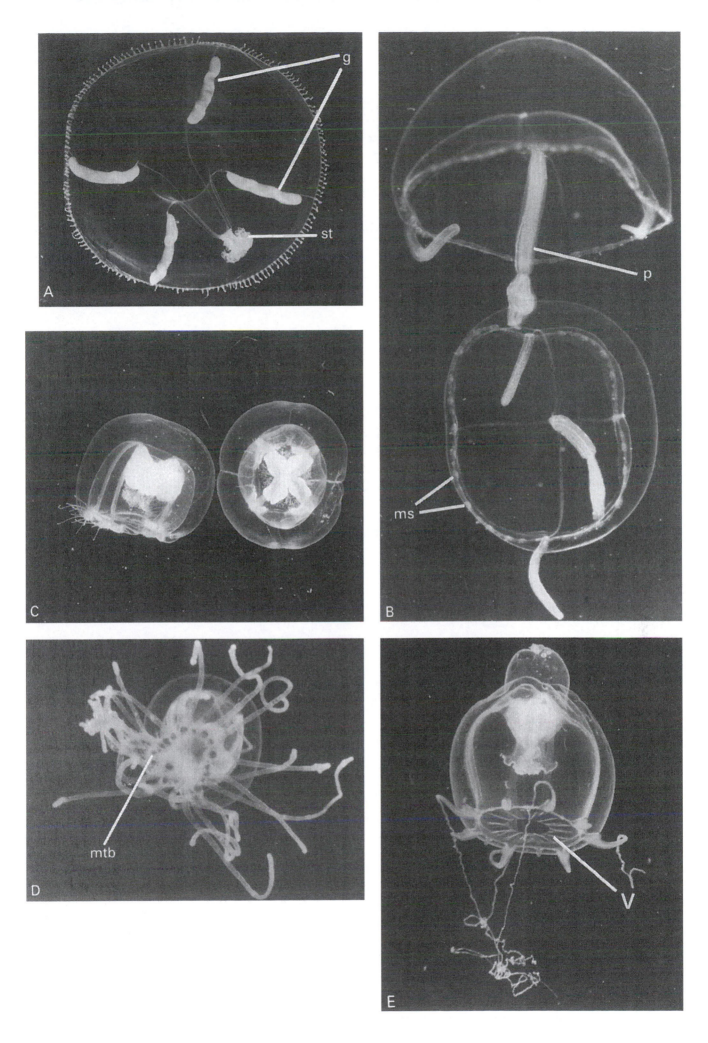

FIGURE 3: A–D Phylum Cnidaria, Class Hydrozoa

This figure illustrates two further species of cnidarian medusae and provides additional photographs of *Leuckartiara* (see Fig. 2E).

FIG. 3A. *Leuckartiara* sp. [height of bell 10 mm]. Note the shape of the apical process on the bell and the extent to which the prey-capturing tentacles may reach away from the bell itself. For their size these medusae are effective swimmers (due in part to the relatively large umbrellar volume and the constricting velum (see Fig. 2E)). The four radial canals are clearly visible.

FIG. 3B. *Leuckartiara octona* (Flemming). [bell height 15 mm]. This is a fully grown specimen which has the full complement of marginal tentacles (16–23 for this species). Note, again, the differing shape of the apical process in comparison to Figs. 2E and 3A.

FIG. 3C. *Staurophora mertensi* Brandt. [diameter 5 cm]. This species may be restricted to the east coast of Scotland and north-east England, and is quite unmistakable. The umbrella is considerably flattened and has the stomach, mouth and four radial canals combined to form a distinctive cross. The marginal tentacles are uniformly short and extremely numerous. The umbrella may exceptionally attain up to 30 cm in diameter.

FIG. 3D. *Sarsia tubulosa* (M. Sars). [bell height 10 mm]. The umbrella of this species is distinctly bell-shaped, being taller than wide, and the jelly is moderately thick. There are four extensible marginal tentacles. The stomach may extend well beyond the bell margin (a feature characteristic of this species), as in the illustration. The mouth is simple and tubular.

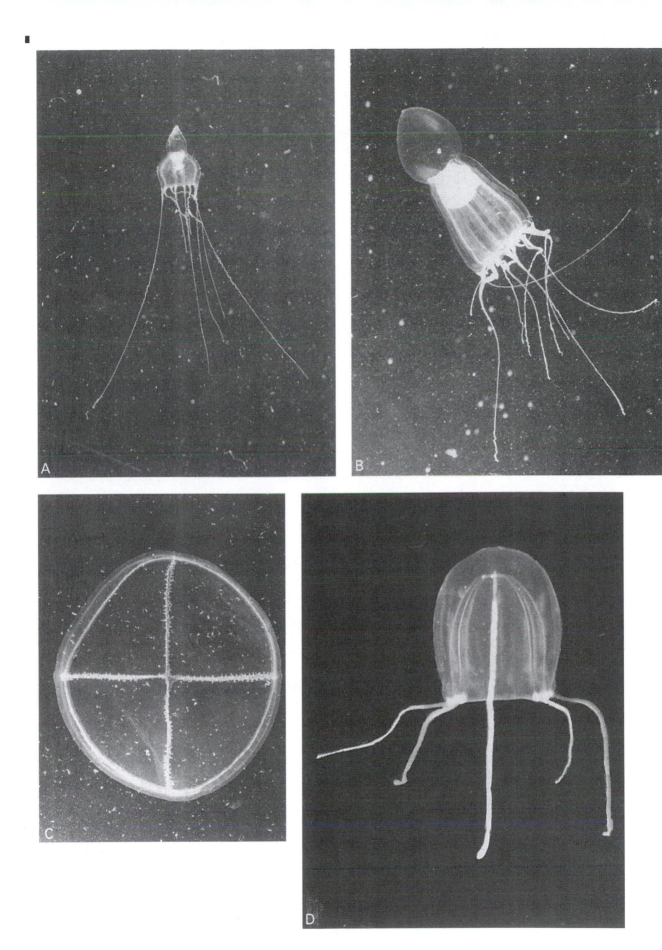

FIG. 4A. *Obelia* sp. [diameter 2 mm]. This small characteristic medusa is that from a species of *Obelia*. Specific diagnosis is problematic, although the medusa of *Obelia longissima* (Pallas) can be recognised by its size (approximately 6 mm diameter) and in having between 100 and 200 marginal tentacles. The presence of the four gonads (g) on the radial canals indicate that the specimen is fully developed and it is therefore likely that this is either *O. geniculata* (L.) or *O. dichotoma* (L.). The umbrella is almost completely flattened, the jelly is thin and there are four simple radial canals. The marginal tentacles differ from many other medusae in being solid and non-retractile.

FIG. 4B. *Cyanea* sp. [diameter 4 mm]. This is the early stage of the ephyra larva of a species of the scyphozoan jellyfish *Cyanea* (see the juvenile in Fig. 5D). In British waters the most frequently encountered ephyra is of *Aurelia aurita* (L.), which can be readily differentiated from *Cyanea* only by examination of the form of the radial canals, the presence of marginal tentacles and the pattern of pigmentation. The ephyra is budded from the benthic polypoid phase of the life cycle (the scyphistoma) by a process of lateral fission called strobilation. A single scyphistoma may bud many (genetically identical) ephyrae during a season. The ephyra – unlike the hydrozoan medusa – can be truly regarded as a larval form because it develops into the distinctly recognisable 'adult' ('medusoid') form before maturing sexually. The scyphozoan ephyra shows remarkable similarity in form between species.

FIG. 4C,D. *Muggiaea atlantica* Cunningham. [length 4 mm]. The order Siphonophora divides into three suborders, the most speciose of which are the Physonectae, e.g. *Agalma*, *Nanomia* (see Fig. 6C), and the Calycophorae. The latter includes the genus *Muggiaea*, illustrated in the present photographs. The calycophorid siphonophores are characterised by two nectophores (or swimming bells), the latero-posterior of which may be reduced or absent. In *M. atlantica* the posterior nectophore is not developed and the single (anterior) nectophore bears distinct ridges which may laterally form a keel (k). The somatocyst (sc) (thought to be a vestigial tentacle) extends the full length of the nectophore and may overtop it (Fig. 4D). The cormidia (c) (secondarily budded polyps) can retract into the space formed by the hydroecium (h). Note the presence of oil-droplets (od) in the somatocyst in Figs. 4C: these may provide some buoyancy, although calycophores are effective swimmers, moving, like medusae, by jet propulsion.

Calycophore siphonophores may, on only cursory inspection, be easily confused with some of the more elongate medusae. Close observation shows the lack of marginal tentacles and of an obvious mouth, but also the presence of the somatocyst. Note the transparency of the colony, which generally persists even in preserved material.

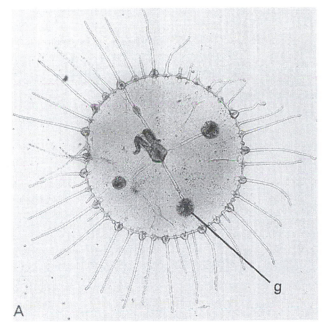

A

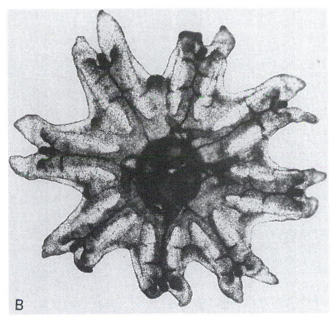

B

g

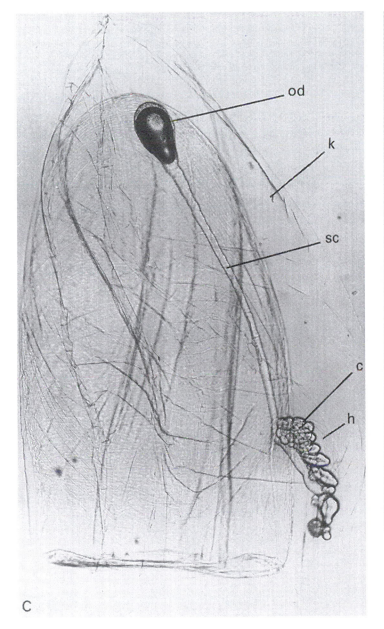

C

od

k

sc

c

h

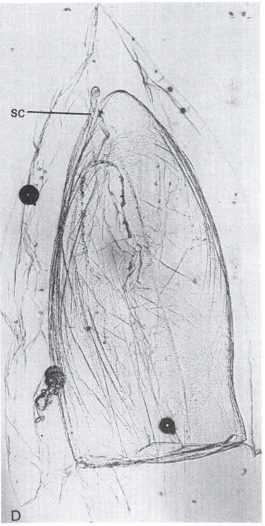

D

sc

II

FIGURE 5: A,B Phylum Ctenophora, Class Tentaculata, Order Cydippida; C Phylum Ctenophora, Class Tentaculata, Order Lobata; D Phylum Cnidaria, Class Scyphozoa; E Phylum Annelida, Class Polychaeta (Family Terebellidae)

FIG. 5A. *Pleurobrachia pileus* (O.F. Müller). [diameter 1 mm]. This small organism is the cydippid larva of the tentaculate ctenophore *Pleurobrachia*. At this stage of development the tentacles are not present, but the characteristic comb plates (cp), or ctenes, are clearly visible. These beat in a metachronal rhythm to provide propulsion.

FIG. 5B. *Pleurobrachia pileus* (O.F. Müller). [body length 15 mm]. [see also Fig. 5A]. This is the fully grown adult form. The rows of comb plates (cp) are clearly seen, as are the pendulous (but highly contractile) tentacles (t) which are totally retractile into sheaths (ts). In contrast to the cnidarians the ctenophores lack nematocysts, but capture and entangle small prey by means of adhesive colloblast cells on the tentacles. Contraction of the tentacles then permits transfer of the prey to the mouth (m). The statocyst (s) is the organ of balance. Note also that the tentacle sheaths (ts) are angled downwards when the organism is in the fishing position; most textbook drawings show ctenophores inverted (with the mouth pointed downwards).

FIG. 5C. *Bolinopsis infundibulum* (O.F. Müller). [length 10 cm]. This is the largest of the British ctenophores. *Bolinopsis* is almost invariably fragmented on capture in plankton nets, because although solid the gelatinous body is extremely fragile. *Bolinopsis* is easily separable from *Beroë* (Fig. 6A,B) by virtue of the large oral lobes (ol) of the former genus: also in *Bolinopsis* the rows of comb plates do not extend the full length of the body. *Bolinopsis* is a voracious predator of cydippid ctenophores.

FIG. 5D. *Cyanea* sp. [bell diameter 2 cm]. (see also Figs. 4B and 8B). This is a juvenile medusa, probably of *Cyanea capillata* (L.). The fully grown medusa may attain almost 1 m across the bell and has tentacles which may extend for several metres. The nematocysts incorporate a virulent toxin which can impart irritable rashes if the medusa is handled. *C. capillata* is invariably brown or red-brown in colour whereas the congener *Cyanea lamarckii* (Péron & Lesueur) is generally blue-purple in coloration and of smaller size (up to approximately 30 cm in diameter). In the illustration the medusa is viewed from above the umbrella: the marginal tentacles (mt) and lappets (lp) – extensions of the subumbrellar margin – are clearly visible.

FIG. 5E. *Lanice conchilega* (Pallas). [length 2 mm]. This shows the post-larval juvenile of the 'sand-mason' worm just prior to settlement. The worm is fully developed and the peristomial palps (pp), by which *Lanice* constructs its sand grain tube and captures food, are visible. Note that the worm – although still pelagic – is occupying a mucous tube (t). After settlement on a suitable sand substratum the juvenile immediately commences cementing sand grains on to the opening of the mucous tube. The tube presumably confers a degree of buoyancy on the developing juvenile because active swimming by this polychaete is not possible. The juveniles are, however, extremely active within the mucous tube while planktonic, and will readily reverse their orientation without actually having to exit the tube itself. Growth of the larva is by budding of new segments at the pygidium (pg), and settlement on to the substratum is probably passive once the larva and tube exceed a critical size.

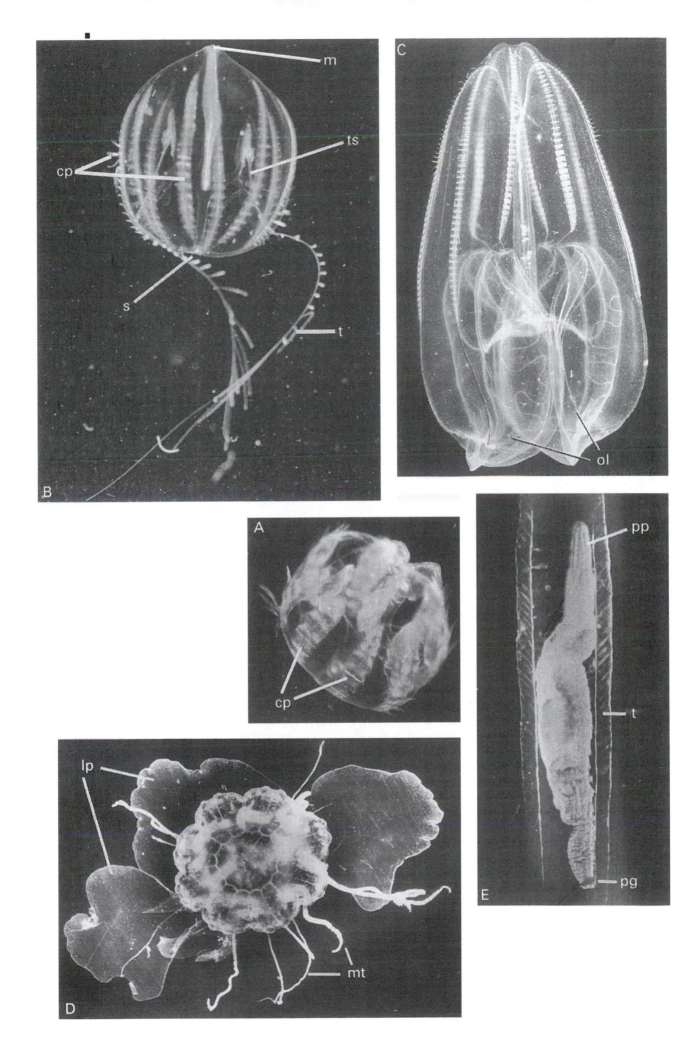

FIGURE 6: A,B Phylum Ctenophora, Class Nuda, Order Beroida; C Phylum Cnidaria, Class Hydrozoa – Order Siphonophora

FIG. 6A. *Beroë cucumis* Fabricius. [length 8 cm]. This is a fully grown specimen. In contrast to *Pleurobrachia* and *Bolinopsis* the gelatinous body of *Beroë* is somewhat opaque: the rows of comb plates often are pink in colour. The position of the mouth (m) is shown, as are the longitudinal rows of comb plates (cp). Note the metachronal rhythm of the arrowed row of comb plates. Despite their size these ctenophores are still strictly planktonic (as opposed to nektonic, e.g. fishes) in that they are largely incapable of swimming against the current. In drifting with the water mass, however, organisms such as these are still capable of orientational swimming and this is effected by the rows of comb plates.

FIG. 6B. *Beroë cucumis* Fabricius. [length 8 cm]. This ctenophore feeds on other comb jellies (especially cydippids) and rapidly engulfs them with its mouth widely dilated as shown in this photograph.

FIG. 6C. 'Stalk' [5 cm] of a physonectid siphonophore colony, probably *Agalma* or *Nanomia*. The siphonophores are specialised colonial hydrozoans which comprise a combination of polypoid (feeding) and modified medusoid (propulsion) individuals. Buoyancy of most siphonophores is provided by the pneumatophore (pn) (or 'float'), which is a gas-filled structure formed by the larval polyp. Siphonophores such as *Agalma* and *Nanomia* are, however, extremely delicate organisms and usually all that the observer sees of them in plankton hauls are the detached swimming bells. In some species (but not *Physalia*) the volume of the pneumatophore can be adjusted, so permitting the colony to sink away from the surface during rough weather.

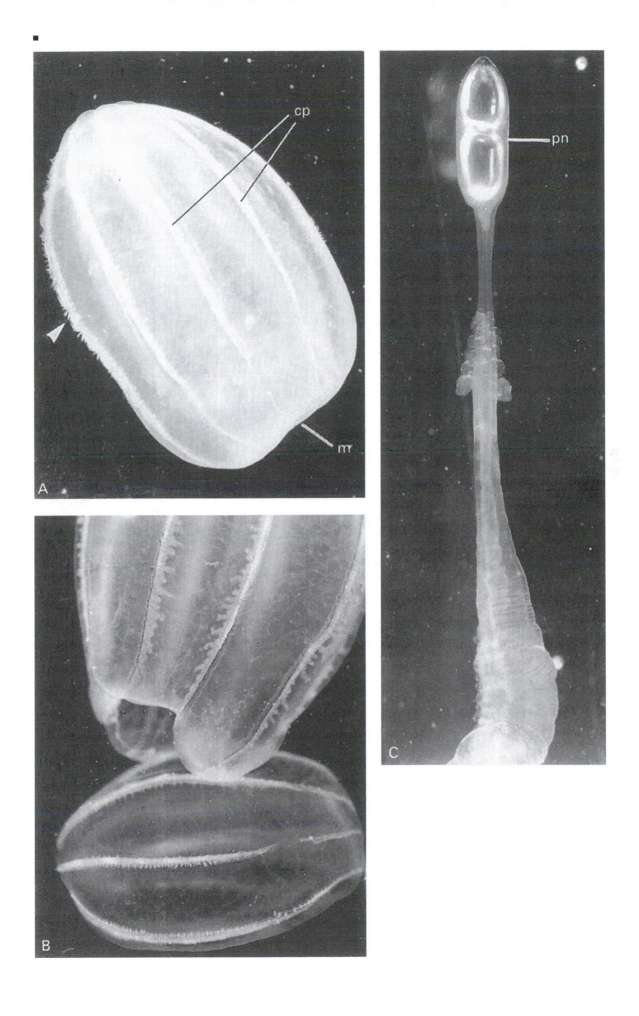

FIGURE 7: A Phylum Nemertea; B Phylum Annelida, Class Polychaeta; C Phylum Annelida, Class Polychaeta (Family Syllidae); D Protozoa, Phylum Sarcomastigophora, Class Dinoflagellida

FIG. 7A. Unidentified nemertean pilidium larva. [diameter 200 μm]. Most nemertines undergo 'direct', or non-pelagic, development and hatch from the egg as fully-formed benthic juveniles. In some species, however, the offspring hatch – as illustrated here – into a free-swimming pelagic larva termed a pilidium. The pilidium has a characteristic form with the body comprising more or less extensive lappets (la) which are marginally ciliated (c). These ciliary bands effect locomotion but the large apical tuft (at) of prominent cilia remains immobile. The apical tuft has a sensory function.

FIG. 7B. Unidentified polychaete trochophore larva. [length 2 mm]. The typical larval form of the polychaete annelids is the trochophore. This is generally spherical or top-shaped and usually has a major 'equatorial' ciliary band (the prototroch (pt)), perhaps in addition to others, according to species. Metamorphosis to the adult vermiform morphology is usually progressive (rather than cataclysmic), with elongation of the body by the budding of segments at the terminal pygidium (pg). This particular specimen is almost certainly the larva of a polynoid 'scaleworm' (see also Fig. 11A): three segments – which already bear stiff chaetae (c) – are visible.

FIG. 7C. *Autolytus edwardsi* (Müller). [length 2.5 mm]. Most polychaetes are benthic as adults yet many have planktonic larvae (e.g. Fig. 7B). A few species are, however, free-swimming and pelagic throughout their life cycle (see *Tomopteris*, Fig. 8E). An unusual life history is displayed by syllid polychaetes (see Fig. 11B-E) in which free-swimming reproductive individuals are budded asexually by the benthic stage. The illustrated female syllid can be seen to be carrying its clutch of eggs. Despite the perhaps unstreamlined appearance, this polychaete swims extraordinarily rapidly by means of lateral flexing of the body, in a typically polychaete fashion. Note the parapodial cirri (pc) and notopodial chaetae (nc), which serve to increase the effective body surface area during swimming.

FIG. 7D. *Noctiluca scintillans* (Macartney). [diameter up to 1 mm]. A very large dinoflagellate (Protozoa) which floats at the surface. Flotation is achieved by means of a large vacuole containing ammonium chloride. The majority of dinoflagellates (e.g. *Ceratium*, *Polykrikos*) are small and often green and claimed by botanists to be plant-like. *Noctiluca* is the largest of the dinoflagellates and carries two flagellate structures, one small and the other stout and long. It is predatory, eating a wide variety of other small plankters, and may occur in swarms of vast numbers, floating as a greasy pink scum on the surface. It is luminescent (hence the specific name), emitting a greenish flash when disturbed.

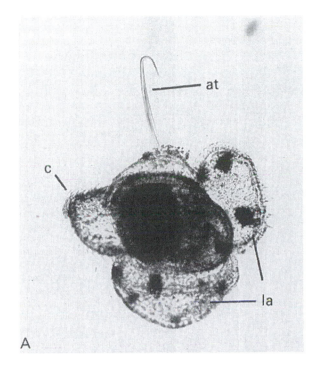

A

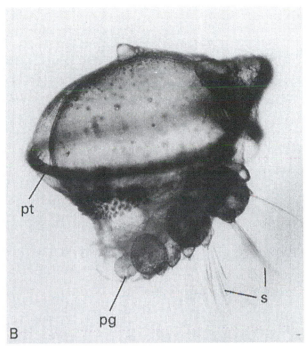

B

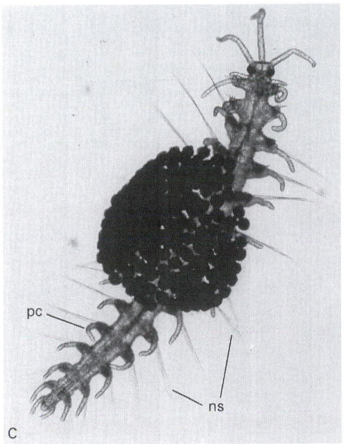

C

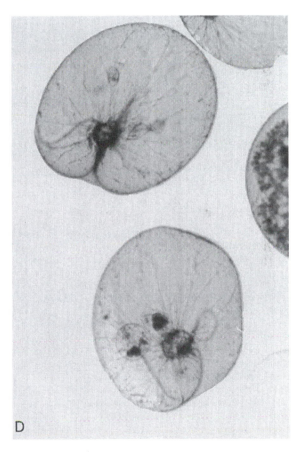

D

FIGURE 8: A Phylum Mollusca, Class Gastropoda, Subclass Opisthobranchia – Order Gymnosomata; B Phylum Cnidaria, Class Scyphozoa; C Phylum Mollusca, Class Gastropoda, Subclass Prosobranchia – Order Mesogastropoda; D Phylum Mollusca, Class Gastropoda, Subclass Prosobranchia – Order Neogastropoda; E Phylum Annelida, Class Polychaeta (Family Tomopteridae)

FIG. 8A. *Clione limacina* Phipps. [length 5 mm]. *Clione* is an example of the shell-less holoplanktonic pteropod molluscs, which swim by means of 'wings' (w), or so-called 'parapodial' extensions of the foot. The latter term is to be avoided if only to preclude confusion with annelid parapodia. The second order of the pteropod molluscs is the Thecosomata (which are shelled), an example of which is illustrated in Fig. 39A,B. *Clione* is, in fact, a specialist predator on its shelled relative *Limacina*. *Clione* swims rather effectively and clasps its prey by means of the conical projections (cp) of the proboscis. It is usually bright pink to orange in colour.

FIG. 8B. *Cyanea* sp. [bell diameter 3 mm]. This is a young post-larval jellyfish of the genus *Cyanea*. The ephyra larva and small medusa of this genus have already been illustrated in Figs. 4B and 5D. In lateral view the large manubrium, or mouth, (m) of this cnidarian can be seen clearly.

FIG. 8C. *Lamellaria perspicua* (L.). [shell length 0.9 mm]. This veliger larva (the so-called echinospira) of this altogether unusual prosobranch gastropod is unique in having a bilaterally flattened *double*-spired transparent shell, the keel of which is ridged and spiny (hence echinospira). The visceral mass (vm) is attached deep within the shell, and in the larva illustrated the eyes (e) can be seen in the typical position at the bases of the velar lobes (v).

FIG. 8D. *Aporrhais pes-pelecani* L. [shell length 2.4 mm]. The adult of this soft-sediment burrowing prosobranch has a characteristic shell with an enlarged wing-like extension to the aperture (hence the specific name 'pelican's foot'). The larval shell (the protoconch) – which is retained following metamorphosis to the benthic adult form – is, however, not yet asymmetrical or flanged. The velum (v) is unusual in being tri-lobed on either side of the head. In this individual the eyes (e), and developing adult 'antennae' (or head tentacles) (a), can be seen in the typical position at the bases of the velar lobes. The anterior portion of the foot (the propodium (pp)) can also be seen to be fully extended.

FIG. 8E. *Tomopteris helgolandica* Greef. [length 3 cm]. This large (up to 5 cm) predatory polychaete is pelagic throughout its life cycle and displays several obvious adaptations to holoplanktonic life: *(i)* the whole organism is transparent, *(ii)* with the exception of the second body segment (which bears antenna-like chaetae), the parapodia (p) lack chaetae, *(iii)* the parapodia bear paddle-like extension plates to enhance swimming ability, *(iv)* it swims exceedingly rapidly. Total transparency is, however, of limited effectiveness as camouflage when one considers the visibility of consumed prey items and gut contents. In the case of *Tomopteris* the filled gut is usually clearly visible (as in the illustration) as a distinct stripe running the length of the body. The rapid swimming of the worm does, none the less, render it extremely difficult to see or follow.

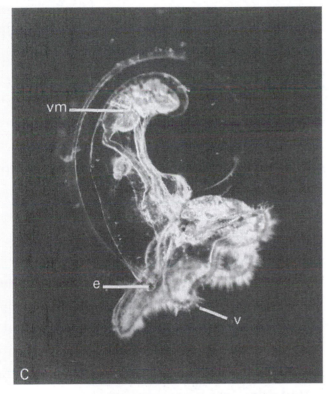

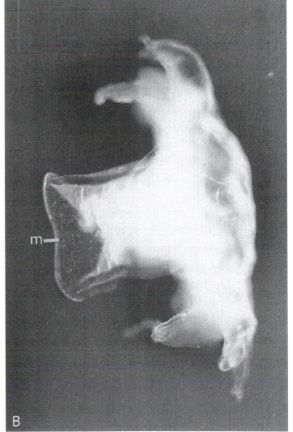

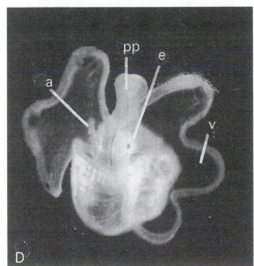

This figure illustrates a number of species of larval polychaetes at varying stages of their development. Note that in one illustrated species (*Pectinaria (=Lagis) koreni* Malmgren, Fig. 9B) the larvae occupy tubes and yet they are still planktonic.

FIG. 9A. *Magelona papillicornis* Müller. [length 1.5 mm]. This is a well-developed individual (cf. Fig. 9E) which is probably still some weeks from settlement: the larva may attain 5 mm before development is complete. This polychaete is characterised by the large grooved peristomial palps (pp) which are muscular, continuously active and which bear obvious papillae (p) (hence the specific name). It should, however, be noted that these palps are readily lost on preservation.

FIG. 9B. *Pectinaria (=Lagis) koreni* Malmgren. [length 2 mm]. These three late larvae are probably within a few days of settlement. The benthic adult *Pectinaria* is sand-dwelling, constructing a short, stiff, conical tube of sand grains. The pelagic juveniles secrete and occupy a mucous tube in a manner similar to that displayed by *Lanice conchilega* Pallas (Fig. 5E): following settlement, *Pectinaria* commences construction of the adult tube by cementing grains to the opening of the larval tube. Note that in contrast to *Lanice* the larval tube of *Pectinaria* is tapered – the larva does not reverse its orientation as does *Lanice*. Close inspection of the tube shows that it comprises longitudinal rows of rhomboid plates: that of *Lanice* is a simple thick tube.

FIG. 9C. *Pectinaria (=Lagis) koreni* Malmgren. [length 600 µm]. This is an early larval stage of those shown in Fig. 9B. This particular larva has developed through the trochophore stage and some 10 or 12 segments have been budded posteriorly from the pygidium (pg) as the larva elongates. Characteristic of this species at this stage of development are the large prototrochal lobes (pl) which project laterally from beneath the prototroch (pc) ciliary girdle itself.

FIG. 9D. *Owenia fusiformis* Delle Chiaje. [body diameter 200 µm]. The unusual form of this trochophore, the so-called 'mitraria' larva which characterises the oweniids, is quite unmistakable. The larval body is shaped like a bishop's mitre (hence the name) and a bundle of numerous elongate chaetae projects from the posterior region. Propulsion of the larva does not involve the chaetae, but is provided by the prototrochal ciliary girdle.

FIG. 9E. *Magelona papillicornis* Müller. [length 1 mm]. This is an early stage of the larva illustrated in Fig. 9A. Note that the palp tentacles, which are striking and characteristic features of the more advanced larvae of this species, have not been lost but are only just commencing development. This species appears to be a specialist predator of bivalve mollusc larvae while planktonic, and a recently captured bivalve veliger (bv) can be seen in the gut.

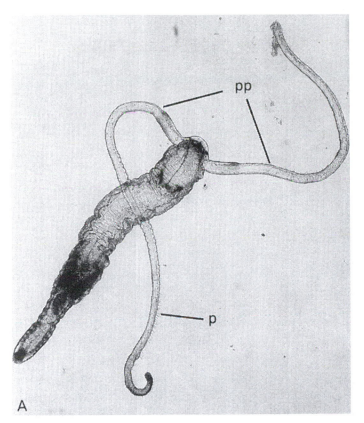

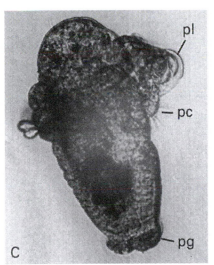

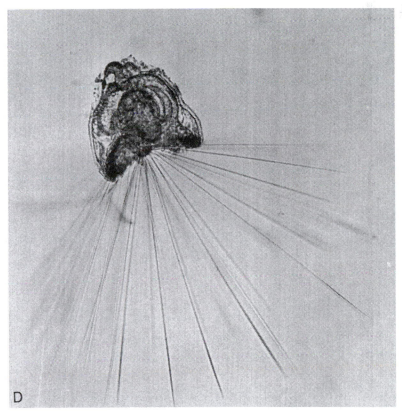

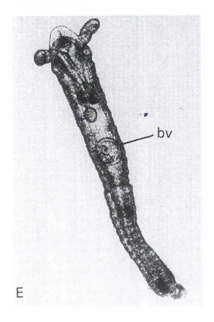

FIGURE 10: A Phylum Annelida, Class Polychaeta (Family Spionidae); B,C Phylum Annelida, Class Polychaeta; D,E Phylum Annelida, Class Polychaeta (Family Tomopteridae)

This figure illustrates three additional larval polychaetes and also shows one of the more prominent holoplanktonic annelids.

FIG. 10A. Unidentified spionid larva, probably *Polydora* sp. [length 600 μm]. Larvae of the family Spionidae are recognisable by their prominent peristomial palps (pp) and the extensive chaetae (c) on the developing segments. The long chaetae, when extended from the body, help to keep the larva suspended in the water column. As the larva grows, the palps develop further and the segments increase to perhaps 20–30 in number, at which time the juvenile settles to the substratum. The illustrated larva has approximately 10 segments.

FIG. 10B,C. Unidentified polychaete larvae. [length 2 mm]. These two specimens are probably close to settlement, having the appearance of benthic juvenile worms. This is suggested by their size, number of segments, and the advanced state of development of the parapodia.

FIG. 10D,E. *Tomopteris helgolandica* Greef. [length 40 mm]. (see also Fig. 8E). This spectacular polychaete is an example of a totally pelagic (holoplanktonic) species of annelid. This is a large, actively swimming predator which displays numerous adaptations to planktonic life, focused especially on maximal camouflage and reduced weight. The whole animal is markedly transparent, the parapodia (p) lack chaetae, but are themselves expanded and bear paddle-like extensions (Fig. 10E). The eversible pharynx, which is used in prey capture, has no teeth or other armament. The long 'antennae' (a) (which are, in fact, the only remaining segmental chaetae) arise from the second segment of the body: the first segment is present in juvenile worms, but absent in the adult. These 'antennae' are probably tactile and chemosensory, as are the prostomial palps (pp), but may function also in buoyancy.

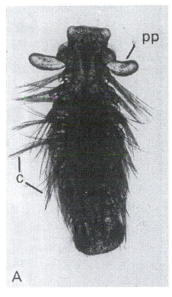

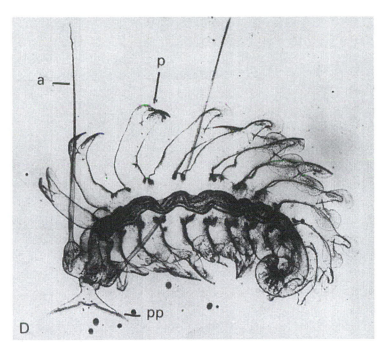

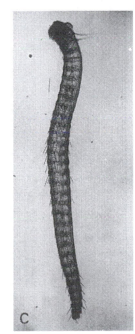

23

FIGURE 11: A Phylum Annelida, Class Polychaeta (Family Aphroditidae); B–E Phylum Annelida, Class Polychaeta (Family Syllidae)

This figure illustrates a well-developed polynoid larva, and also the stolons of several syllid polychaetes of the genera *Autolytus* and *Proceraea*. The life cycle of *Autolytus* is complex, and varies in detail between species. The essential feature, however, is that there is a benthic predatory 'stock', incapable of swimming, from which reproductive (non-feeding) 'stolons' are budded asexually. Each stock buds either only male or only female stolons which swim up into the plankton. Sexual reproduction occurs during this pelagic phase, which then gives rise to further benthic individuals.

FIG. 11A. *Harmothoë imbricata* L. [length 2.5 mm]. This is a fully developed juvenile which is ready to settle to the substratum. Eight heavily chaetigerous body segments are apparent, and the elytra (e) – notopodial plates which cover the dorsum – can clearly be seen.

FIG. 11B. *Proceraea cornuta* (Rathke). [length 5 mm]. Female syllid stolons, such as the present individual, are characterised by having simple head antennal appendages (ha). This female is seen in lateral view. The eggs are matured within the body cavity and then extruded as a spawn mass (sm) (contained within a membrane) which is retained by the adult (see also Fig. 7C).

FIG. 11C. *Autolytus edwardsi* (Müller). [length 2.5 mm]. This female stolon is in what is known as the post-hatching state. That this is a female is indicated by the simple filiform head appendages (ha), but the spawn mass is missing. (Pre-spawning females can be recognised from the large numbers of eggs still held within the body cavity.) The pre-hatching state is illustrated in Fig. 7C.

FIG. 11D. *Autolytus brachycephalus* (Claparéde). [length 2.5 mm]. This individual clearly is a male, as shown by the bifid head antennae (bha) and the heavily muscular swimming parapodia (sp). The three dorsal head 'horns', or palps (pp), which are characteristic of *Autolytus* and *Proceraea* males, are not clearly visible in this illustration, having been curled back and displaced to the right. Diagnostic of the male of this species are the three segments in the 'front' region (fr) of the stolon (anterior to the 'middle' (swimming) region (mr) and the 'tail' region (not shown)).

FIG. 11E. *Proceraea prismatica* (Fabricius). [length 5 mm]. The bifid head antennae (bha), elongate (but contracted) dorsal head 'horns', or palps (pp), and large swimming parapodia (sp) confirm this individual to be a male. The three regions of the stolon – the 'front' (fr), 'middle' (mr) (swimming) and 'tail' (tr) – are clearly visible. Note also the large eyes (e), which can have no predatory function because the stolon is non-feeding: it is likely that they are behaviourally important in the location of females or in vertical migration.

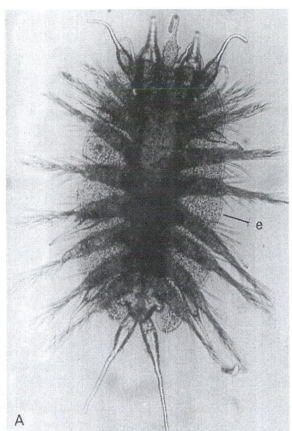

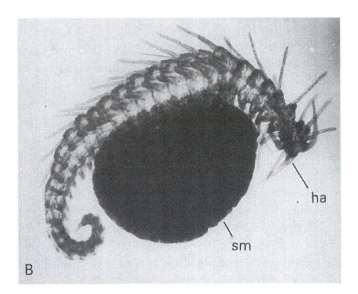

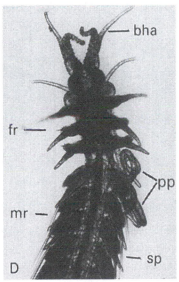

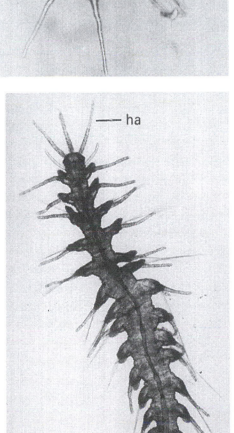

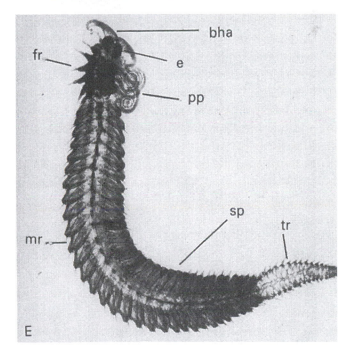

FIGURE 12: Phylum Arthropoda, Class Crustacea, Subclass Copepoda – Order Calanoida

Calanoid copepods are, numerically, the dominant organisms amongst zooplankton communities around the British Isles, although at varying localities and in different seasons of the year other taxa may predominate. It is important, therefore, that the various species of copepods should be distinguished.

Planktonic copepods are almost invariably species of the order Calanoida, although some species of the orders Cyclopoida and Harpacticoida may be encountered. Copepods may be considered to be configured by (*i*) five head segments, (*ii*) seven thoracic segments, and (*iii*) four abdominal segments including the telson bearing the paired caudal rami. The head/thorax segments each carry paired appendages in the sequence: *Head* (antennule, antenna, mandible, maxillule, maxilla) and *Thorax* (maxilliped, thoracic swimming limbs 1–5 or 1–6). The abdominal segments bear no appendages. For all copepods the first thoracic segment at least is fused with the head segments and covered dorsally by the carapace to form the cephalosome. From three to five free thoracic segments remain (according to species) and these, plus the cephalothorax, comprise the prosome in calanoids. In females, segments 1 and 2 of the urosome are fused to form the genital double segment (or genital complex; see Fig. 20): typically, therefore, females have a shorter urosome of fewer segments than do the males. The telson bears the anus and a pair of caudal rami which carry more or less elaborate setae.

Copepodids – immature post-naupliar stages – are normally difficult (if not impossible) to identify and can be recognised as such by their having fewer than the complete five pairs of thoracic swimming limbs or an incomplete series of abdominal segments. Some taxonomic authorities use the term copepodite as an alternative to copepodid.

Cyclopoid copepods have the articulation between prosome and urosome, not between the sixth thoracic and genital segments (as in calanoids), but between the fifth and sixth thoracic segments. This gives cyclopoids the appearance of a long slender urosome and the antennules are usually shorter than is typical of calanoids. Harpacticoids tend to have a continuous slender body profile, with no obvious articulation distinguishing prosome from urosome. The antennules also are generally very short and bear numerous setae, and the caudal rami often bear extremely long setae.

Most reproductive female calanoid copepods (cf. *Caligus elongatus* Nordmann; Fig. 20), do not retain eggs in external egg sacs but release them to the water column. The nauplius larval stage is succeeded by a series of five copepodid stages, CI–CV, each separated by a moult, before the adult status is achieved.

FIG. 12.. *Calanus helgolandicus* (Claus). [length 3.5 mm]. Antennule (the first cephalic appendage) (a1), antenna (a2) (the second cephalic appendage), mandible (ma), cephalosome (c), prosome (ps), urosome (us), caudal rami (cr), caudal setae (cs), free thoracic segments (fts), genital segment (gs), telson (t), lipid deposits (ld), articulation of cephalosome with the first free thoracic segment (arrowed). (See also Fig. 13A.)

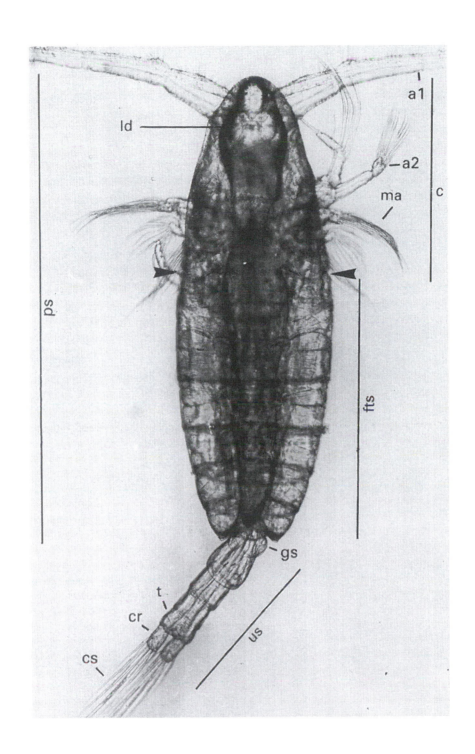

FIGURE 13: A–C Phylum Arthropoda, Class Crustacea, Subclass Copepoda – Order Calanoida

This figure illustrates one of the most easily recognisable and widespread species of calanoid copepod around the British Isles (*Calanus helgolandicus* (Claus)), and another (*Metridia lucens* Boeck) for which *Calanus* is often mistaken.

FIG. 13A. *Calanus helgolandicus* (Claus). [length 3.5 mm]. (see also Fig. 12). This intermediate-sized copepod is extremely difficult to distinguish from *Calanus finmarchicus* (Gunnerus). The genus *Calanus* is characterised by the beautifully streamlined torpedo-shaped prosome, five free thoracic segments, and extremely long antennules which bear three large setae (two backward- and one forward-pointing) at their tips (s). In length the antennules slightly exceed that of the entire body. The urosome is very short, as are the caudal rami themselves: each ramus bears long caudal setae, although these are often broken in specimens taken in plankton nets.

FIG. 13B,C. *Metridia lucens* Boeck. [length 2 mm]. This moderately large calanoid is superficially similar to *Calanus*, but in fact is readily distinguishable in having only four free thoracic segments, rather short antennules, and a long urosome. *M. lucens* is colourless, almost pellucid, in appearance and is luminescent (hence the specific name).

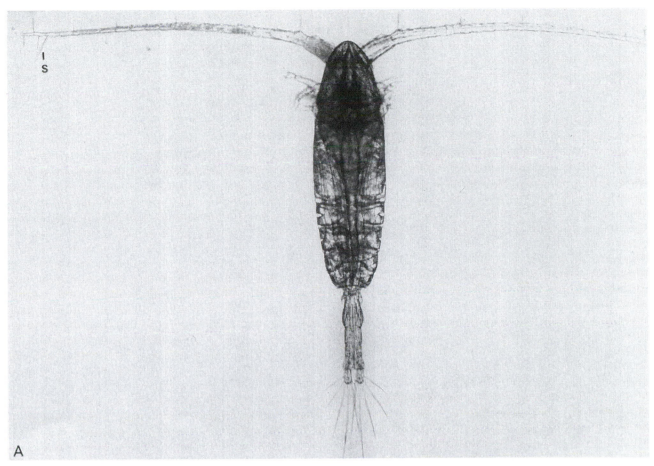

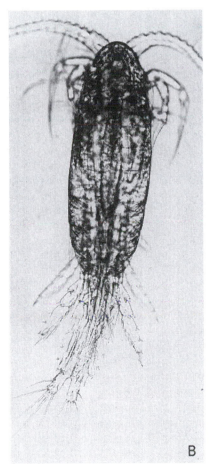

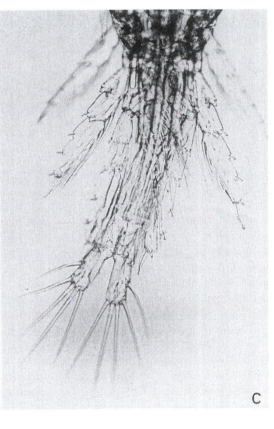

29

FIGURE 14: A–E Phylum Arthropoda, Class Crustacea, Subclass Copepoda – Order Calanoida

There are three common species of small calanoids in the size range of 0.7–1.5 mm, all of which have only the three free thoracic segments and which are broadly similar in appearance. By estimation of the relative lengths of the urosome versus the prosome, the absolute length of the antennules, by counting the number of urosome segments, and by examination of the antennulary setae not only are these species separable but individuals may also be sexed.

FIG. 14A. *Pseudocalanus elongatus* Boeck. [female, length 1.3 mm]. The three free thoracic segments are obvious. The urosome (excluding the caudal setae) is exactly half the length of the prosome. The head has a characteristically smooth and streamlined dorsal profile (cf. *Microcalanus*, Fig. 14B). The genital double segment (gs) is larger than the other urosome segments.

FIG. 14B,C. *Microcalanus pusillus* Sars. [male, length 0.7 mm]. The urosome equals half the length of the prosome in the male (it is shorter in the female). The antennules are short and do not reach to the posterior margin of the prosome (see Fig. 14C). In the male the basal segments of the antennules are fused, the setae are enlarged and there are additional chemosensory setae (aesthetascs) present (the same is seen in *Pseudocalanus elongatus* Boeck and *Paracalanus parvus* (Claus) males). The head

forms a characteristically triangular point when viewed dorsally (Fig. 14B – cf. *Pseudocalanus*).

FIG. 14D. *Paracalanus parvus* (Claus). [length 1 mm]. This individual is the copepodid IV stage. Typically, for this species, the cephalosome is twice the length of the (three) free thoracic segments. The urosome is approximately $1/3$ the length of the prosome, and comprises three segments at this stage. The antennules are about as long as the entire body.

FIG. 14E. *Paracalanus parvus* (Claus). [length 1 mm]. This individual is probably the male copepodid V (pre-adult) stage. The urosome consists of four segments at this stage, the last of which will subdivide at the final moult to produce the five-segmented urosome of the adult male.

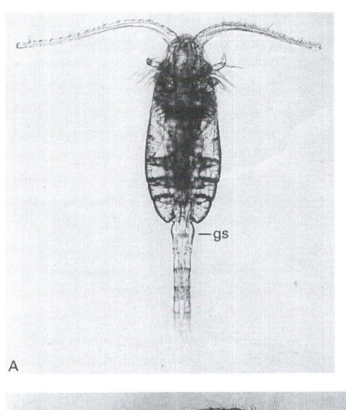

A

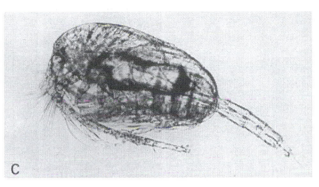

C

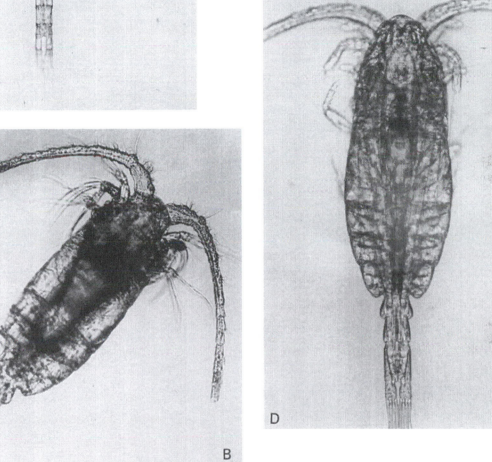

D

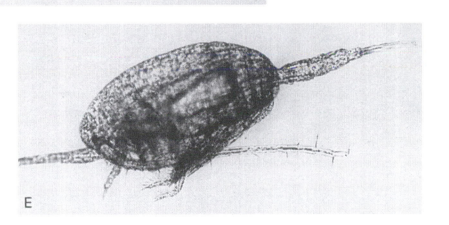

B

E

31

FIGURE 15: A–E Phylum Arthropoda, Class Crustacea, Subclass Copepoda – Order Calanoida

FIG. 15A, B. *Eurytemora hirundoides* (Nordqvist). [length 1.1 mm]. This is a female, which still bears a few eggs on the urosome. The genera *Eurytemora* and *Pseudocalanus*, and the family Euchaetidae, are unusual amongst calanoids in carrying their fertilized eggs in external egg sacs (see Fig. 12). *Eurytemora* is characterised by its small size, five free thoracic segments (the fifth of which in the female is fluted into lateral points (lp)) and very elongated caudal rami (cr) which are approximately 10x as long as broad. The antennules are short and do not extend to the end of the prosome; note the long scale-like spermatophore (sp), which is deposited on the female by the male, emanating from the genital segment. *Eurytemora* is characteristic of brackish estuaries, where it is tolerant of low salinity, high temperature and low oxygen tension.

FIG. 15C. *Temora longicornis* (O.F. Müller). [copepodid, length 1.0 mm]. (see also Fig. 16D,E). Note the absence of the lateral points on the last free thoracic segment (cf. *Eurytemora*) and the short three-segmented urosome at this stage. A well-defined genital segment is lacking because the individual is not mature.

FIG. 15D,E. *Pareuchaeta norvegica* (Boeck). [length 7.5 mm]. This very large, bright crimson copepod is quite unmistakable. There are only four obvious free thoracic segments. The head is pointed and the heavily setose antennules are very typical, but diagnostic is the appearance of the caudal setae (cs) on the caudal rami (cr). The long setae are almost feather-like in appearance and quite unlike any other calanoid. The female of *Pareuchaeta norvegica* is predatory (as evidenced by the large and elaborate head appendages (mandibles, maxillules, maxillae, maxillipeds)) and often spends extended periods hanging still, but upside down, in the water column: males are non-feeding and have atrophied mouthparts. Like most calanoid copepods, and despite its size, *Pareuchaeta norvegica* can swim extremely rapidly. Around the British Isles *Pareuchaeta norvegica* is especially common in some of the deep sea lochs (fjords) of the Scottish west coast.

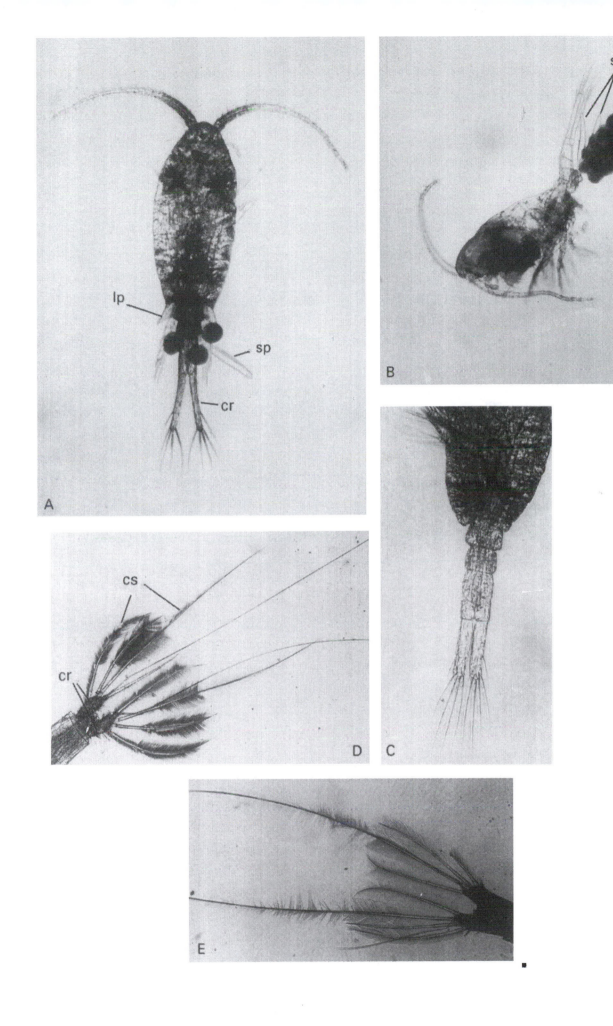

FIGURE 16: A–E Phylum Arthropoda, Class Crustacea, Subclass Copepoda – Order Calanoida

This figure illustrates two common genera of calanoid copepod (*Acartia* and *Temora*) which are quite different in appearance, and which are easily recognisable. The various species of *Acartia* (viz. *Acartia clausi* Giesbrecht, *A. discaudata* Giesbrecht, *A. bifilosa* (Giesbrecht), and *A. longiremis* (Lilljeborg)) are difficult to distinguish and the appropriate specialist taxonomic works should be consulted.

FIG. 16A,B. *Acartia clausi* Giesbrecht. [length 1.1 mm]. There are four free thoracic segments, the first of which is larger than the other three, but whose articulation with the cephalosome is often difficult to see (arrowed in the figure). The antennules are slightly shorter than the prosome and are heavily, but unevenly, setose. The setae on the caudal rami (Fig. 16B) characteristically are hand-like, being five in number and spread like fingers. The prosome has a typically streamlined, bullet-shaped, form with a blunted (but none the less slightly pointed) head.

FIG. 16C. *Acartia clausi* Giesbrecht. [length 1.1 mm]. The female has rows of small points (p) along the posterior margin of the dorsal surface of the urosome segments.

FIG. 16D,E. *Temora longicornis* (O.F. Müller). [female, length 1.5 mm]. This unmistakable copepod has a distinctly pear-shaped prosome, with a broad cephalosome and four free thoracic segments. These features, combined with the long caudal rami (Fig. 16E) – approximately 5x as long as broad – place this species without confusion. The female has a urosome of only three segments while the male has a urosome of five segments.

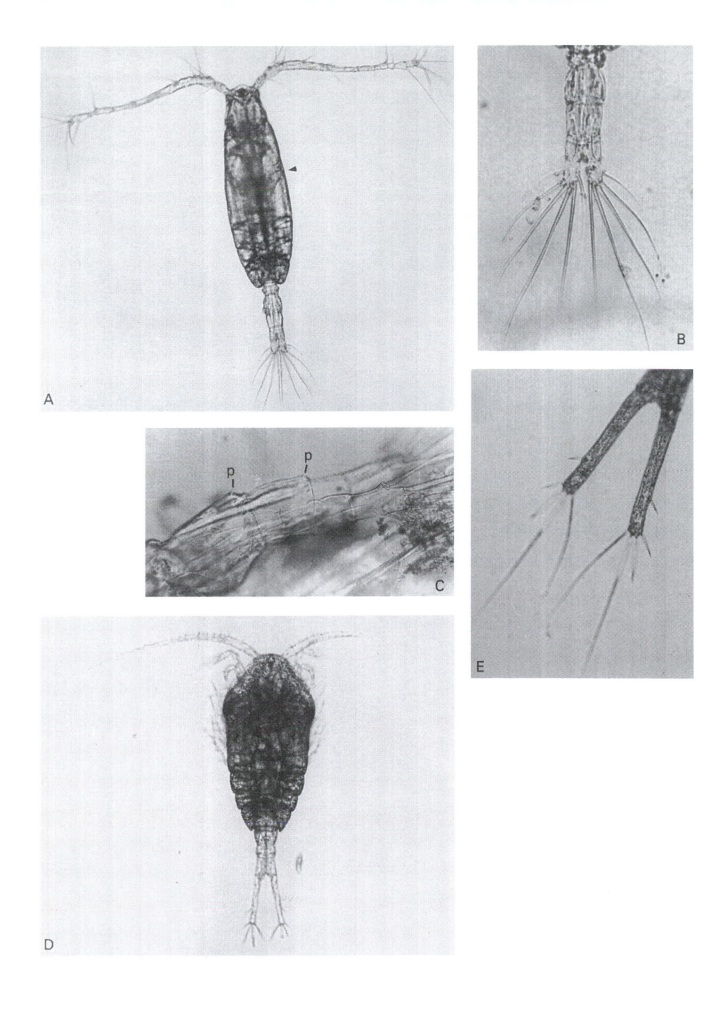

FIGURE 17: A–F Phylum Arthropoda, Class Crustacea, Subclass Copepoda – Order Calanoida

FIG. 17A,C. *Centropages typicus* Krøyer. [male, length 1.7 mm]. Characteristically, for *Centropages*, there are five free thoracic segments and the anterior region of the cephalosome has a slightly 'pinched-in' appearance although the head is still rather pointed (cf. *Candacia*, Figs. 17F, 18D). Also the fifth free thoracic segment is laterally produced into spines (s). The male is recognised by having four urosome segments, whereas there are three in the female. The right limb of the sixth pair of thoracic appendages, the fifth pair of legs, is modified into a clasping chela (ch). In *C. typicus* the thoracic spines are more or less symmetrical. Note also the modified prehensile right antennule (pa) (for grasping the female).

FIG. 17B. *Centropages typicus* Krøyer. [female, length 1.7 mm]. (cf. Fig. 17A,C). The female of this species is characterised by the symmetrical spine points (s) of the fifth free thoracic segment, the three urosome segments, and the simple antennules.

FIG. 17D,E. *Centropages hamatus* (Lilljeborg). [female, length 1.4 mm]. Note the asymmetry of the spines (s) on the fifth free thoracic segment. This asymmetry is, however, variable and examples of both the left (D) and the right (E) are illustrated. The three urosome segments (plus caudal rami) show the individual in Fig. 17E to be an adult female. The male of this species, like *Centropages typicus* (Fig. 17A,C), has the right limb of the fifth pair of legs developed into a clasping chela. *C. hamatus* is readily distinguishable from *C. typicus* in being smaller, more slender, having a less swollen right antennule (males only) and a smaller chela (right limb of fifth swimming leg, males only).

FIG. 17F. *Candacia armata* (Boeck). [female, length 2.3 mm]. This genus is similar to *Centropages* but is distinguishable in having only four free thoracic segments, a very square head and somewhat more recurved spines on the fourth free thoracic segment. *Candacia* also is readily recognisable by virtue of the black pigmentation of the setae of the head and thoracic limbs, a feature not lost on preservation.

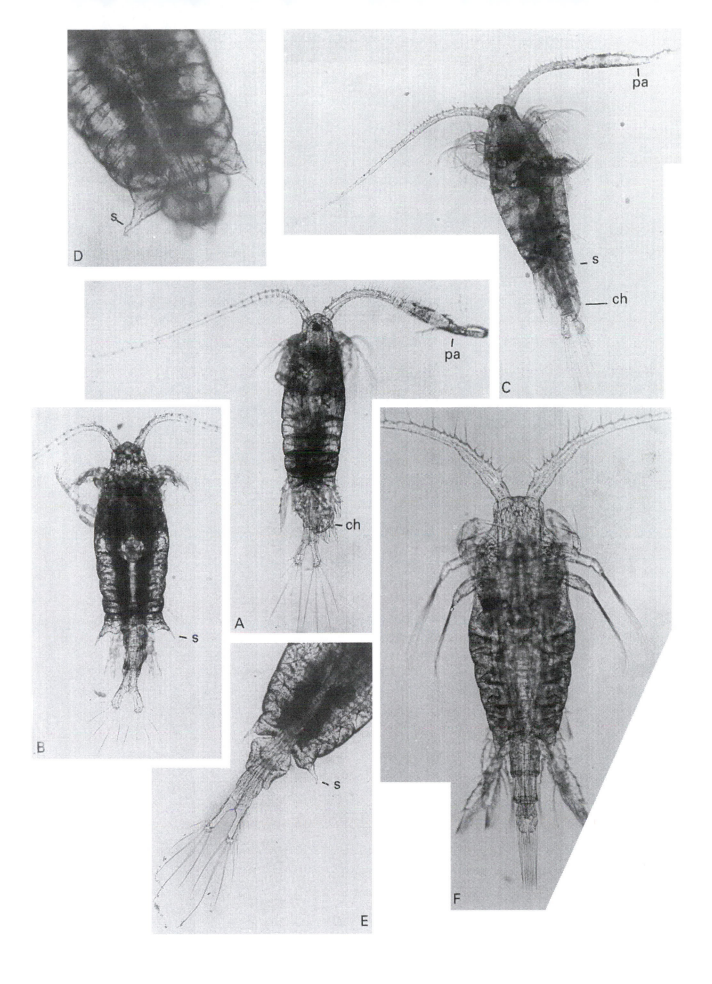

37

FIGURE 18: A–D Phylum Arthropoda, Class Crustacea, Subclass Copepoda – Order Calanoida

This figure illustrates the two pontellid calanoid copepods likely to be encountered in plankton samples and also shows the male of *Candacia armata* Boeck (to be compared with the female shown in Fig. 17F). The pontellids *Anomalocera* and *Labidocera* are unmistakable calanoids in view of their moderately large size and bright coloration. Their blue, or blue-green, coloration is indicative that they are primarily surface-dwelling: blue is the predominant colour among marine pleuston organisms. Pleuston is the collective term applied to animals living at the sea–air interface: typical examples include the cnidarian chondrophores (e.g. *Velella*) and the siphonophore *Physalia*. Plankton refers to those (invariably small) organisms which are compelled to drift with the current, whereas nekton includes those larger forms (such as fish and cephalopods) which can actively and directionally swim against the current.

FIG. 18A,B. *Labidocera wollastoni* (Lubbock). [female, length 2.3 mm]. This species is yellowish to bluish-green in colour and has moderately long antennules which reach to perhaps half-way along the urosome (antennule tip arrowed in Fig. 18A). The prosome tapers to the anterior and posterior ends and the anterior cephalosome bears two pairs of lateral hooks (lh) (posterior to the eye (e)) resulting in an arrowhead-like shape when viewed dorsally. The last free thoracic segment is produced into two symmetrical points (p). There are five free thoracic segments. The urosome of the female consists of only two segments plus the telson: the genital segment is quite swollen in appearance. In the male the urosome has four slender segments plus the telson, and the right antennule is markedly swollen and prehensile.

FIG. 18C. *Anomalocera pattersoni* Templeton. [copepodid V, male, length 3.0 mm]. This species is similar to *Labidocera* in size and coloration although *Anomalocera* is a much brighter blue, with additional irregular dark patches of pigment which can be seen on the prosome. The last of the five free thoracic segments is drawn out to points (p) of which the right-hand one is slightly asymmetrically in-curved. As for *Labidocera* the anterior cephalosome is drawn out into lateral hooks (lh) to form an arrowhead-shape. The antennules are comparatively short, being only three quarters of the length of the prosome; in the copepodid V male (as illustrated) the right antennule is prehensile (pa), with swollen segments which become much more pronounced in the adult male.

FIG. 18D. *Candacia armata* Boeck. [male, length 2.7 mm] (cf. female in Fig. 17F). The male of *C. armata* can be recognised in having the point of the last free thoracic segment slightly incurved on the right-hand side (p), and by the genital segment of the urosome – this bears a large backward-pointing hooked projection (hp) on the right-hand side.

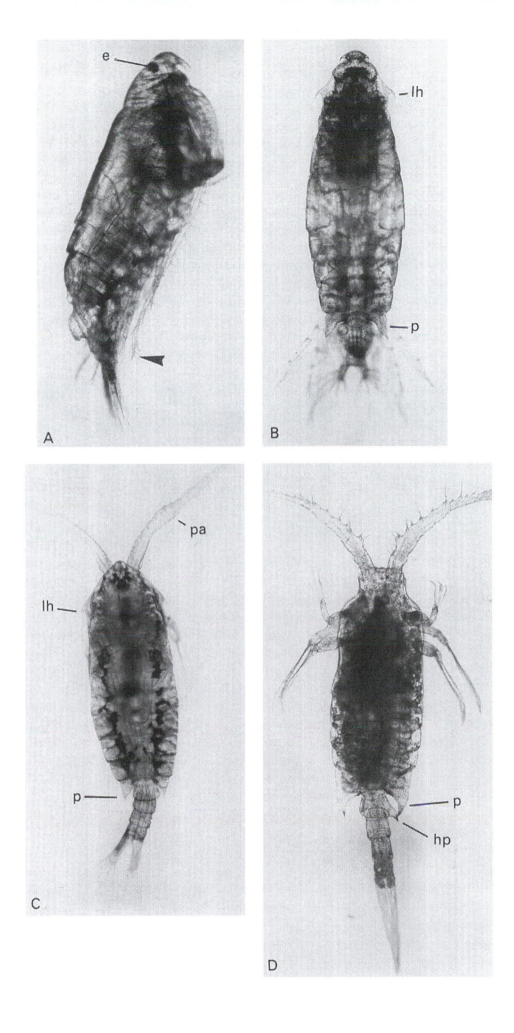

FIGURE 19: A,B Phylum Arthropoda, Class Crustacea, Subclass Copepoda – Order Cyclopoida; C Phylum Arthropoda, Class Crustacea, Subclass Malacostraca – Order Isopoda; D–F Phylum Arthropoda, Class Crustacea, Subclass Copepoda – Order Harpacticoida

FIG. 19A,B. *Oithona helgolandica* Claus. [female, length 0.8 mm]. This cyclopoid copepod has a characteristic appearance, with a streamlined prosome bearing four free thoracic segments, and short but heavily setose antennules (a). The first urosome segment (the sixth thoracic segment) is somewhat rounded and carries two obvious lateral spines on each side: these are not, however, visible in the photographs because of the underlying thoracic swimming limbs. The second urosome (= genital double) segment is elongate and pear-shaped. Three other segments make up the urosome. The caudal rami (cr) are short, but widely divergent, and each bears two major and four minor setae (cs). The thoracic swimming limbs (tsl) bear unusually long setae. The maxillipeds (ma) are especially long for a copepod of this small size. *Oithona* is undoubtedly carnivorous, but is also reported to be omnivorous.

FIG. 19C. Unidentified isopod microniscus stage. [length 0.7 mm]. This is the microniscus stage of an unidentified epicaridean isopod. The microniscus is typically an external parasite of planktonic copepods, including *Calanus*, and can be quite abundant in British waters. Typically, for an isopod, the body is dorsoventrally flattened and there are no obvious specialisations of the body segments into functional regions (tagmata). The antennules (a1) are short and heavily setose, and the antennae (a2) are elongate and setose.

FIG. 19D. *Zaus* sp. [female, length 0.5 mm]. This is an epibenthic harpacticoid which has relatively large antennules (a1) and a shield-like cephalothorax (cx). The free thoracic segments are equally broad, but the abdomen somewhat foreshortened. The caudal rami are small, but the caudal setae (cs) typically stout. These copepods occasionally venture up into the water column.

FIG. 19E,F. *Tigriopus* sp. [early copepodid stages, length 1.2 mm]. Like *Zaus*, this species is of the family Harpacticidae, and yet *Tigriopus* has a more elongate form. These are early copepodid stages. The antennules (a1) are very short but the cephalothorax (cx) is still quite broad. The caudal rami (cr) are very small, despite the large caudal setae (cs).

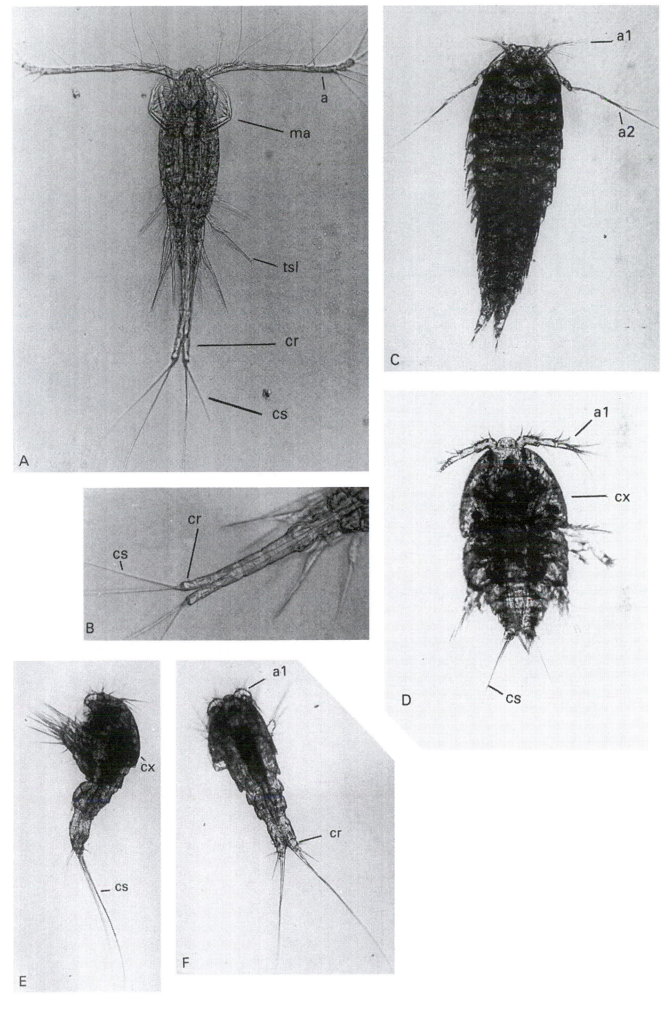

FIGURE 20: Phylum Arthropoda, Class Crustacea, Subclass Copepoda – Order Siphonostomatoida (Family Caligidae)

Many parasitic copepod species are associated with fish hosts and display considerable morphological specialisation in being adapted to this particular mode of life. Indeed, the adults of many species of siphonostomatoids (such as those of the genus *Lernaeocera*, which are gill parasites) are barely recognisable as arthropods, let alone copepods. The larval stages of most parasitic copepods are, however, little different from their free-living counterparts: this is to be expected because the individual requires to be free-swimming as a juvenile in order to infect the (mobile) host fish.

Species of the siphonostomatoid family Caligidae show marked cephalisation, to the extent that only the fifth thoracic segment remains free of the cephalothorax. The cephalothorax is formed of the fused head segments plus the first four thoracic segments, and is almost circular in shape. Posterior to the free thoracic segment is the genital complex, which comprises the fused sixth thoracic and genital segments. In females the oviduct openings are the points of attachment for the egg strings. The abdomen is short and formed of up to three segments (according to species), and is terminally bisected by the anus, on either side of which are the short caudal rami bearing spiny setae.

FIG. 20. *Caligus elongatus* Nordmann. [female, length 5 mm]. The anterior margin of the cephalothorax is characterised by the pair of lunules (l), or attachment suckers, adjacent to the short antennules (a1). The single (fifth) free thoracic segment (ts) bears the only long limbs. The paired egg strings (es) are seen attached to the genital complex (gc), anterior to the abdomen (a).

C. elongatus is the most common of the British caligids, and has been recorded from up to 80 species of fish, including both teleosts (e.g. Cod, Haddock, Ling, Pollack, Bass, Mackerel, Herring, Gurnard, Salmon, Dab, Plaice and Flounder) and elasmobranchs (e.g. Skates, Rays and Dogfish). This species, like other caligids, is strictly ectoparasitic – feeding upon mucous secretions and epidermal tissues of the host fish. Adult caligids do not attach permanently to the host, but frequently move actively over their skin. Actively swimming caligids readily and effectively attach to hard substrata or other large organisms following contact. Attachment of the adult is largely effected by the grasping limbs and the suction provided by the form of the carapace: this has the appearance of an inverted saucer on the skin of the host. Caligids are not especially unusual in plankton samples, and they can be seen to swim extremely effectively, but presumably most caught in plankton nets have become inadvertently separated from their host. The related species *Lepeophtheirus salmonis* (Krøyer) has become a major pest species on commercial marine salmon (*Salmo salar* L.) farms.

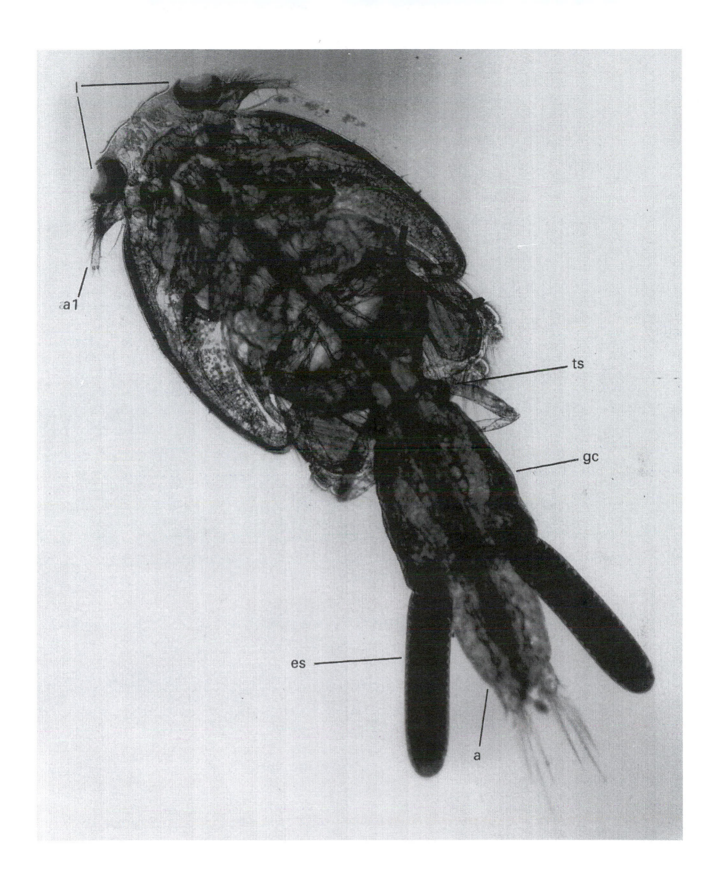

FIGURE 21: Phylum Arthropoda, Class Crustacea

This figure illustrates a variety of crustacean nauplii. All such nauplii are very small, in the region of 100–200 μm in length. The nauplius is characteristically the first post-hatching larval stage of most crustaceans. Amongst the more advanced Crustacea, however, such as the Decapoda, the nauplius is rarely present: the higher Malacostraca hatch at a more advanced stage. As for the barnacle (cirripede) nauplius (in the succeeding illustration) the simple body plan of a carapace and only three pairs of limbs (antennule, antenna, mandible respectively) can be seen.

Larval crustaceans can increase in size and number/complexity of appendages only as they moult. Moreover, there is invariably a marked polarity of development, with the head and thorax appendages/segments being formed before the abdomen. This pattern of development is most clearly seen in malacostracan larvae, in which the head and thorax appendages/segments are well developed before the abdominal pleopods have even formed. Crustacean nauplii, such as those illustrated, are generally difficult to identify with any certainty. However, of the present examples, Fig. 21C can be identified as the nauplius stage of a caligid siphonostomatoid, an adult of which is shown in Fig. 20. Fig. 21D probably is the lecithotrophic (= yolk rich) nauplius of *Pareuchaeta norvegica* (Boeck) (see also Fig. 15D,E) and Fig. 21E can be identified as the nauplius of the harpacticoid genus *Longipedia* Claus.

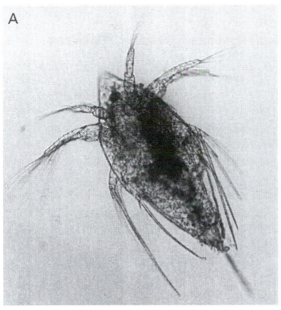

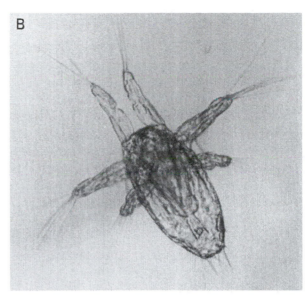

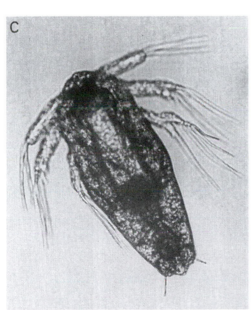

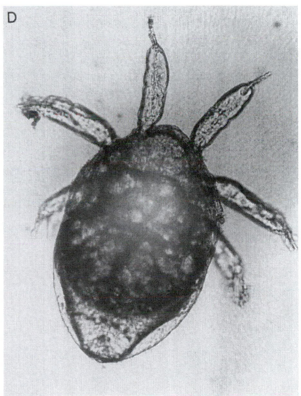

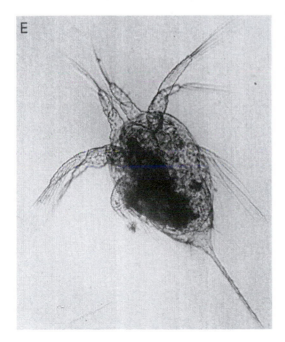

This figure illustrates the two major larval stages in the development of barnacles. The larva hatches as a nauplius, which is characterised by three pairs of limbs (antennules (a1), antennae (a2) and the mandibles (m)). In the nauplius the antennules and antennae have a swimming function, in contrast to their sensory function in the majority of adult crustaceans. In the specific case of the barnacles the post-larval antennules perform yet another function in cementing the individual to the substratum. The single-eyed (e) nauplius proceeds through five moults (six nauplius stages) before moulting into the settling cyprid stage. The cyprid is so-called because of its similarity to an ostracod and is the non-feeding larval stage which metamorphoses into the benthic barnacle.

FIG. 22A. *Semibalanus balanoides* (L.). [length 250 μm]. Nauplius. Barnacle nauplii can be easily recognised and separated from other crustacean nauplii by the presence of the paired fronto-lateral horns (lh) on the carapace. The shape of the ventral thoracic process (tp) is species-diagnostic. The frontal filaments (ff) are shown on each side of the nauplius eye (e).

FIG. 22B,C. *Semibalanus balanoides* (L). [length 600 μm]. The cyprid is a non-feeding final larval stage which undergoes the process of settlement and attachment to the substratum and metamorphosis to the adult form. Being non-feeding the larva has to subsist on stored resources accreted during the naupliar stages – lipid deposits (ld) are particularly notable at the anterior end. The body and appendages are completely enclosed in the hinged bivalved carapace: generally only the antennules, part of the thorax and extremities of the limbs are seen (Fig. 22B), but in Fig. 22C the thorax (t) has been deliberately extended for illustrative purposes. The thorax bears the six pairs of (swimming) thoracic limbs. The adult barnacle feeds by filtering water-borne particles and phytoplankton with the thoracic limbs (cirri, see Fig. 23), which are enclosed within the calcareous shell plates.

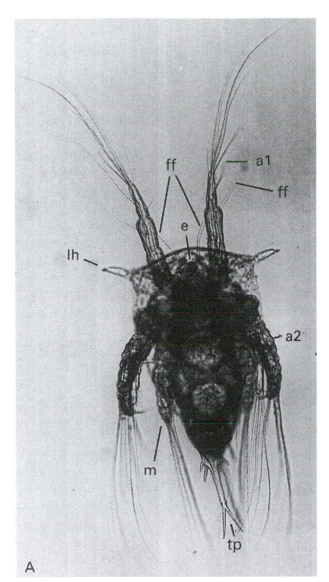

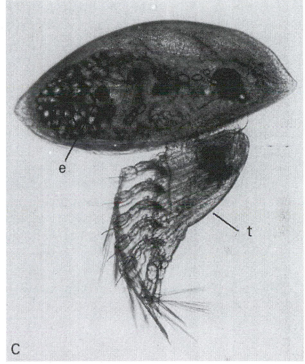

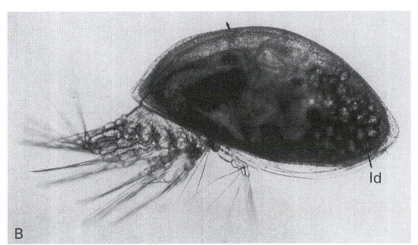

FIGURE 23: Phylum Arthropoda, Class Crustacea, Subclass Thecostraca – Superorder Cirripedia

Plankton samples inevitably contain many incomplete or damaged specimens and this may render unequivocal identification extremely difficult, if not impossible. Net samples do, however, also frequently contain material which is not only difficult to recognise but also not necessarily planktonic. Thus, plastic fragments, paper fibres, drowned terrestrial insects (blown offshore), seed husks, etc. may all be found. Furthermore, should the sampling net touch bottom in shallow water it is likely that some benthic invertebrates may also be caught. Even live broken fragments of truly planktonic organisms may be difficult to recognise at first. Thus, the disrupted tentacles of medusae and comb plates of ctenophores may continue to move for some hours following capture, and may cause considerable frustration in attempting to identify them.

The illustrated 'organism' in this figure is common throughout the spring and summer months in inshore plankton samples, and is regularly attributed to almost every invertebrate phylum. They vary in length and commonly exceed 10 mm. Intact specimens show three pairs of obvious (biramous) limbs (cirri) attached to a wrinkled body. They are totally transparent and lack any evidence of a gut.

This is, in fact, the moult (or exuvium) of an adult barnacle and comprises the posterior three pairs of cirri: the full complement is six pairs, as seen in the cyprid larvae illustrated in Fig. 22B,C. Crustacea, like all arthropods, periodically moult and the exuvium is separated from the body. In being cast away the lightly calcified moult therefore assumes a planktonic existence until it eventually disintegrates. One has only to recall the number of barnacles on rocky shores to appreciate the number of exuviae which must be released to the water column in a year, so it is not surprising that a few are periodically captured.

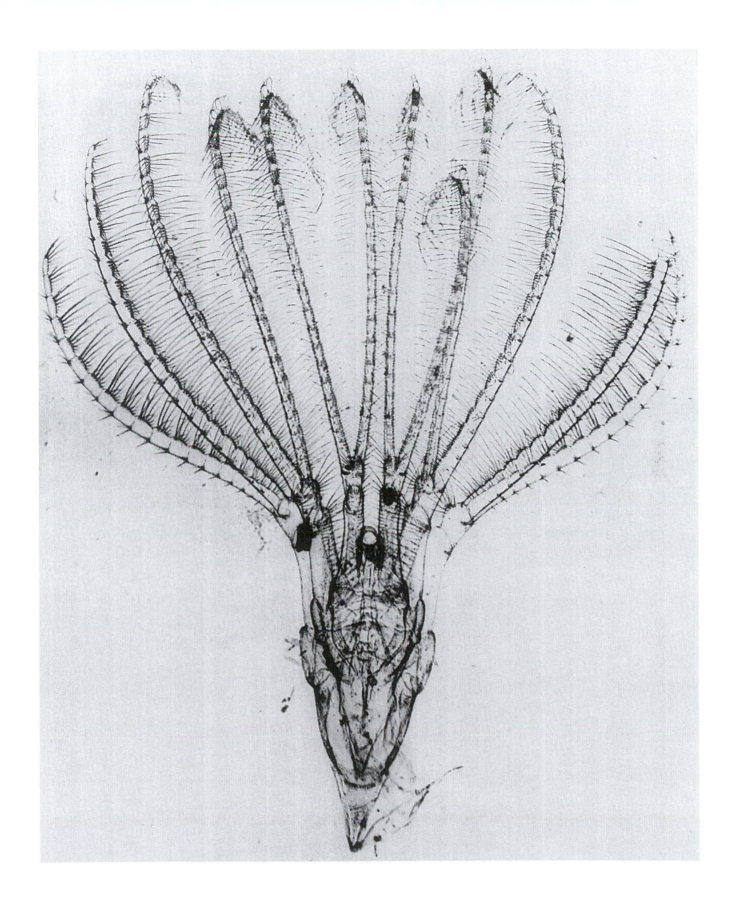

FIGURE 24: A Phylum Arthropoda, Class Crustacea, Subclass Malacostraca – Order Amphipoda; B,C Phylum Arthropoda, Class Crustacea, Subclass Branchiopoda – Order Onchypoda; D Phylum Arthropoda, Class Crustacea, Subclass Malacostraca – Order Decapoda (Section Macrura).

FIG. 24A. *Hyperia galba* (Montagu). [length 13 mm]. The antennules and antennae (a1, a2) are very short and located between the enormous compound eyes. As is typical of amphipods, the thorax lacks a carapace. Despite being generally laterally flattened, the anterior body of this species is considerably swollen when viewed dorsally. The abdomen, bearing the large setose swimming pleopods (p), is comparatively slender. The anterior thoracic walking limbs are sub-chelate for grasping prey items. The large compound eyes (ce) clearly show the predatory habit of this crustacean.

FIG. 24B. *Podon polyphemoides* Leuckart. [length 600 μm]. This characteristic small cladoceran is easily recognisable by the large bulbous compound eye (ce) and the upwardly pointing setose antennae (a2) by which it swims. The antennules are extremely small and barely visible.

FIG. 24C. *Evadne nordmanni* Loven. [length 600 um]. This cladoceran has a more elongated appearance than *Podon*, but also shows the typical cladoceran compound eye. Note the large numbers of developing embryos (e) within the brood cavity. *Evadne* is distinguishable from *Podon* by the terminal point (p) to the carapace and more streamlined body. Reproduction here is parthenogenetic and results in rapid local increases in numbers. Such seasonal population 'explosions' may temporarily exclude other zooplankters, on a local scale.

FIG. 24D. *Upogebia deltaura* (Leach). [length 5mm]. (see also Fig. 29C). This is the late larva of a mud-burrowing prawn-like macruran malacostracan. In this species the abdomen is not asymmetrical (cf. hermit crabs, see page 62, Fig. 30) and is characterised by the well-developed pleopods (p), the large fan-like telson (t), the heavily setose chelipeds (ch) and walking limbs, and the unusual chelae – in which the movable joint (the dactylus) is longer than the immovable part of the pincer (the propodus).

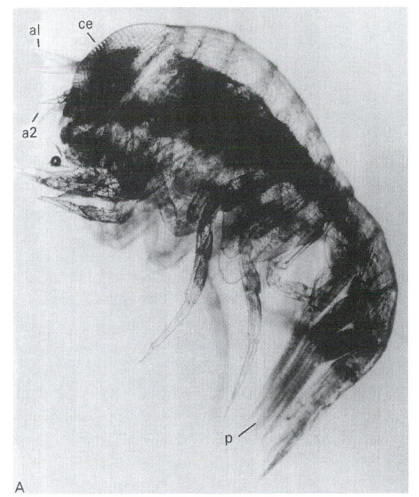

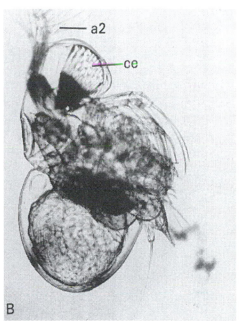

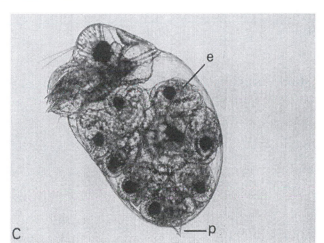

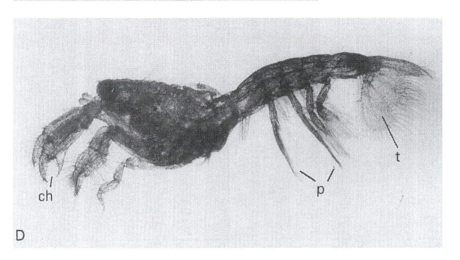

FIGURE 25: A–F Phylum Arthropoda, Class Crustacea, Subclass Malacostraca – Order Mysidacea

Mysids are shrimp-like crustaceans often found in dense aggregations. Many are epibenthic, being active swimmers but nevertheless associated with the substratum. The carapace is large, but unlike the superficially similar Euphausiacea (see Fig. 26), and the higher Decapoda, the mysid carapace covers (but does not fuse with) the thorax. Mysids have stalked eyes, and the thoracic limbs bear long exopods which operate rhythmically during swimming and filter-feeding. The female carries the eggs within a brood-pouch (hence the colloquial term 'opposum shrimps') composed of oostegites on up to seven thoracic limbs. Nearly all mysids have a pair of statocysts located within the uropods. Unlike the euphausiids, mysids do not have photophores.

FIG. 25A–F. *Praunus flexuosus* (Müller). [length 2 cm]. Fig. 25A shows the anterior end in dorsal view. Here the large stalked eyes, the expansive antennal scales (as) and the filiform thoracic limbs are all evident. External ciliate protozoan infestations (cil, Fig. 25B) are not uncommon on the carapace of *Praunus* from shallow water, but do not appear to harm the crustacean. The small abdominal pleopods (pl), which lie against the ventral surface, can be seen in Fig. 25C.

Mysids have the ability to change colour very rapidly and can match a wide array of backgrounds. Fig. 25D shows the large ventral abdominal chromatophores (ch), and Fig. 25E the uropod chromatophores, in this species. Fig. 25E also shows the large statocysts (st) located within the uropods of the tail-fan. Note the feather-like setae of the uropod margins. Fig. 25F shows the tail-fan and last three abdominal segments in dorsal view.

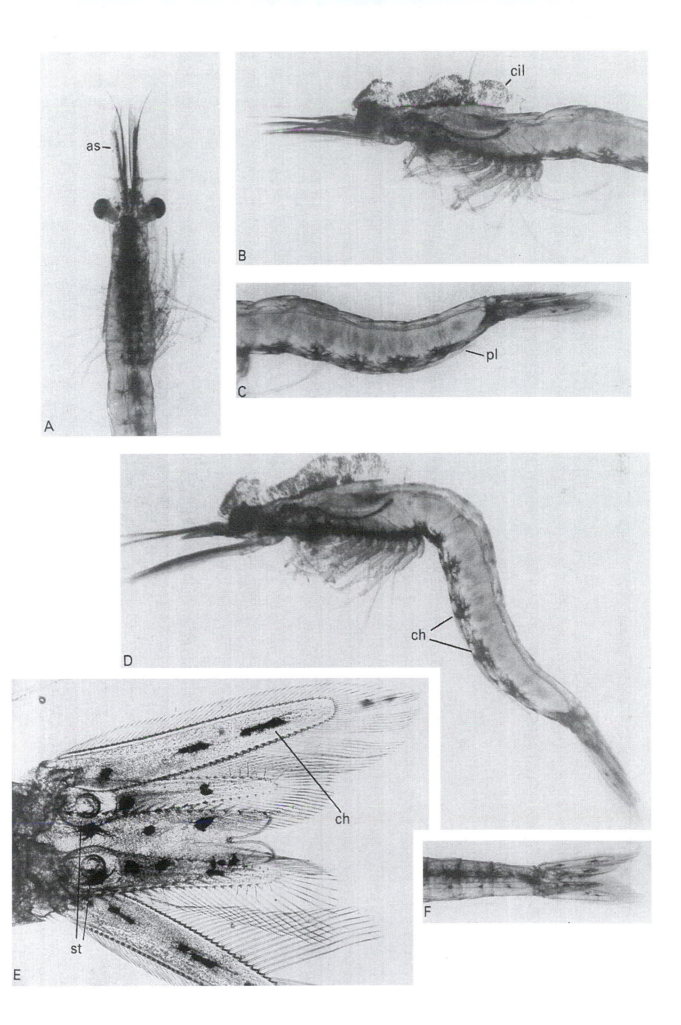

FIGURE 26: A–C Phylum Arthropoda, Class Crustacea, Subclass Malacostraca – Order Euphausiacea

The Euphausiacea are a group of shrimp-like crustaceans very similar at first glance to mysids (see Fig. 25). The euphausiids are usually larger, and the carapace is fused to the thoracic segments. Euphausiids readily can be distinguished from mysids by *(i)* the absence of uropod statocysts (see Fig. 25E,F), *(ii)* the presence of photophores (p) at the abdominal pleopod (pl) and the thoracic (tl) limb bases, and *(iii)* the presence of feathery gills (g) – modified epipodites – at the bases of the thoracic limbs. The fundamental body plan of the malacostracan crustaceans is the subject of some controversy, but it can be argued that it is one of five or six head, eight thoracic, and six or seven abdominal segments, with their respective appendage pairs. The sixth abdominal segment has a double nerve ganglion (hence the differing interpretations of six or seven segments) and the first, or preantennulary, head segment (if actually present) does not bear appendages. The sequential segmental appendages include, respectively, the antennules, antennae, mandibles, maxillules, maxillae, up to eight thoracic 'walking' limbs, and five abdominal pleopods: the first pair of pleopods is modified into a copulatory petasma in males and the final abdominal segment(s) do not bear appendages. In most malacostracans up to three anterior pairs of thoracic legs are modified as maxillipeds. The thoracic limbs of euphausiids are similar and all are biramous, with the exopodites active and rhythmic in life: no thoracic limbs are modified into maxillipeds as they are in mysids. Euphausiids are all filter-feeders and comprise the pivotal taxon ('krill') of the pelagic ecosystem in the Southern Ocean.

FIG. 26A–C. *Meganyctiphanes norvegicus* (M. Sars). [length 20 mm]. In this species the eighth pair of the thoracic limbs (tl) is rudimentary. This species may attain 4 cm in length. *M. norvegicus* lacks a rostral spine between the compound eyes (ce), but the antennules (a1) and antennae (a2) are large. The setose antennal scales (as) are particularly well developed.

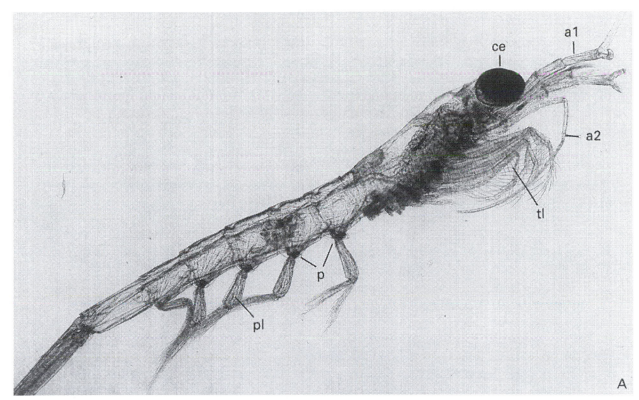

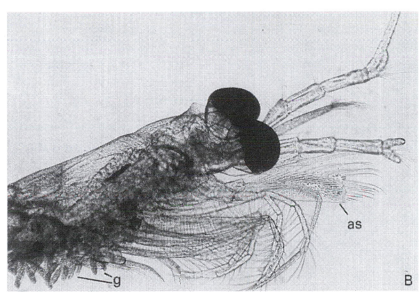

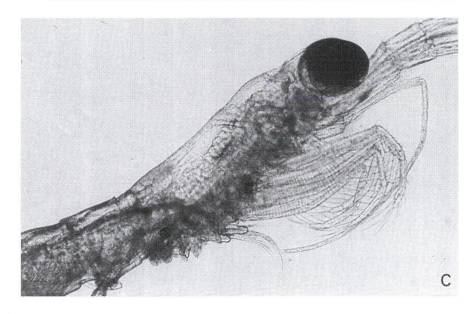

FIGURE 27: A,B Phylum Arthropoda, Class Crustacea, Subclass Malacostraca – Order Euphausiacea; C Phylum Arthropoda, Class Crustacea, Subclass Malacostraca – Order Decapoda (Section Macrura); D,E Phylum Arthropoda, Class Crustacea, Subclass Malacostraca – Order Decapoda (Section Caridea)

This figure illustrates a variety of malacostracan larvae. Note that although they display a superficial similarity, close examination shows clear differences which permit ready identification not only to group but often to species.

FIG. 27A,B. *Nyctiphanes couchii* (Bell). [length 3 mm]. This is the stage 3 furcilia larva. Euphausiids hatch from the egg as a non-feeding nauplius stage and this moults through a succession of calyptopis (protozoea) > furcilia (zoea) > cyrtopia (post-larval) stages in progressing to the adult form. Note the antennules (a1) and antennae (a2) differentiate early and that the abdomen consists already of the definitive six segments, but as yet bears no obvious pleopods. The uropods (u) and telson (t), and their arrangement of spines (Fig. 27B), are species-diagnostic of these euphausiids.

FIG. 27C. *Axius stirhynchus* Leach. [length 2 mm]. This is a well developed, but not yet fully formed, juvenile: note that the pleopods (pl) are still bud-like. The telson, on which the anus (a) can be seen, is quite unmistakable. This larva is almost totally transparent and has a glass-like appearance. The rostral spine (r) is long and prominent, as are the antennal scales (as).

FIG. 27D. Unidentified hippolytid larva. [length 3.0 mm]. This caridean larva is probably a stage 2 zoea of a species of the family Hippolytidae (see also Fig. 28F, which probably is the zoea 1 of the same species). The uropods (u) are not especially well developed at this stage but, again, the size and shape of the abdominal segments, and the telson (t) in particular, aid identification.

FIG. 27E. Unidentified hippolytid larva. [length 3.5 mm]. This figure shows the stage 3 or 4 zoea of probably the same species as Fig. 27D. This lateral view reveals the thoracic limbs (tl), the abdominal pleopods (pl) and the relative positions of the uropods (u) and telson (t). Note that the lengths of the uropods, relative to the telson, are somewhat greater than for the individual in Fig. 27D, showing this to be a later zoeal stage.

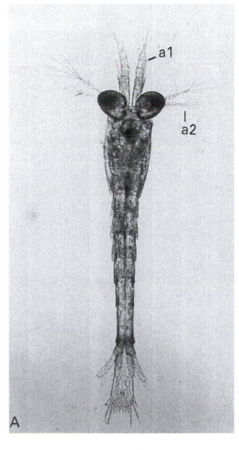

A

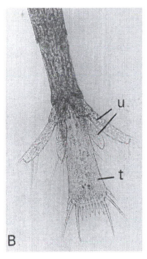

B

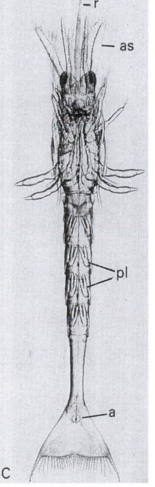

C

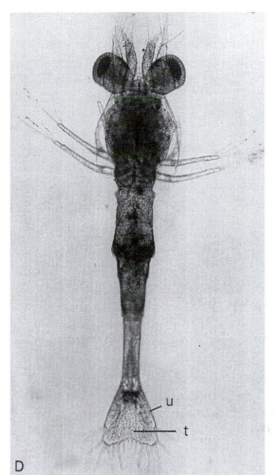

D

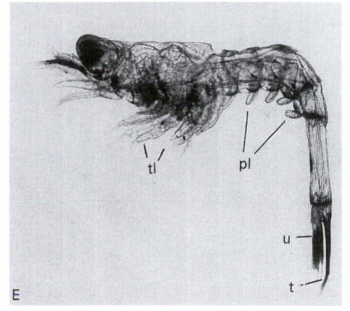

E

FIGURE 28: A,B,C,F,G Phylum Arthropoda, Class Crustacea, Subclass Malacostraca – Order Decapoda (Section Caridea); D,E Phylum Arthropoda, Class Crustacea, Subclass Malacostraca – Order Decapoda (Section Macrura)

This figure illustrates different developmental stages of a variety of caridean and macruran larvae. As with most crustacean larvae, the eyes, head appendages and telson/uropods prove to be the most characteristic structures to examine in attempting to identify these organisms.

FIG. 28A. *Philocheras* sp. [length 3.5 mm]. This figure illustrates an early zoea stage, because the uropods (u) have not yet erupted. This possibly is *Philocheras echinulatus* (M. Sars) and has a strikingly spinulated telson, a long 'neck' (n) of the telson itself, and a pair of obvious spines (s) on the previous abdominal segment. The genus *Philocheras* is of the family Crangonidae, which includes the common brown shrimp *Crangon crangon* (L.)

FIG. 28B,C. Unidentified crangonid larva. [length 2 mm]. This caridean larva probably is of the family Crangonidae and has a characteristic telson. Note the obvious spines on the telson margin and the large chromatophores (c). It is a later larval stage than that illustrated in Fig. 28A, for example, because the uropods are well developed and functional.

FIG. 28D. *Callianassa subterranea* (Montagu). [length 4 mm]. This is the stage 2 zoea; as in Fig. 28A this larva has yet to erupt the uropods. The spinulation of the telson is quite unmistakable. Note the large, broad, serrated rostral spine (r) and the characteristic elongated compound eyes (e).

FIG. 28E. *Callianassa subterranea* (Montagu). [length 4.5 mm]. This is the larval stage succeeding that shown in the previous illustration, following the moult. (In being restrained for photography, the carapace of this individual has been somewhat distorted.) Note that the uropods (u) have now erupted and that the thoracic limbs are more obvious. None the less, there is, as yet, no development of the abdominal pleopods.

FIG. 28F. Unidentified hippolytid larva. [length 2 mm]. This probably is the first larval stage (zoea 1) of a shrimp of the family Hippolytidae (see also the later zoea stages in Fig. 27D,E). The telson (which lacks free uropods) has a distinctive form and spinulation.

FIG. 28G. *Processa nouveli holthuisi* Al-Adhub & Williamson. [length 5 mm]. This is the final, stage 6, zoea larva. Note the considerable development of the antennules (a1), antennae (a2) and the antennal scales (as). Note also the difference in appearance of the telson and the size of the uropods (u) compared to the stage 5 zoea (Fig. 29B).

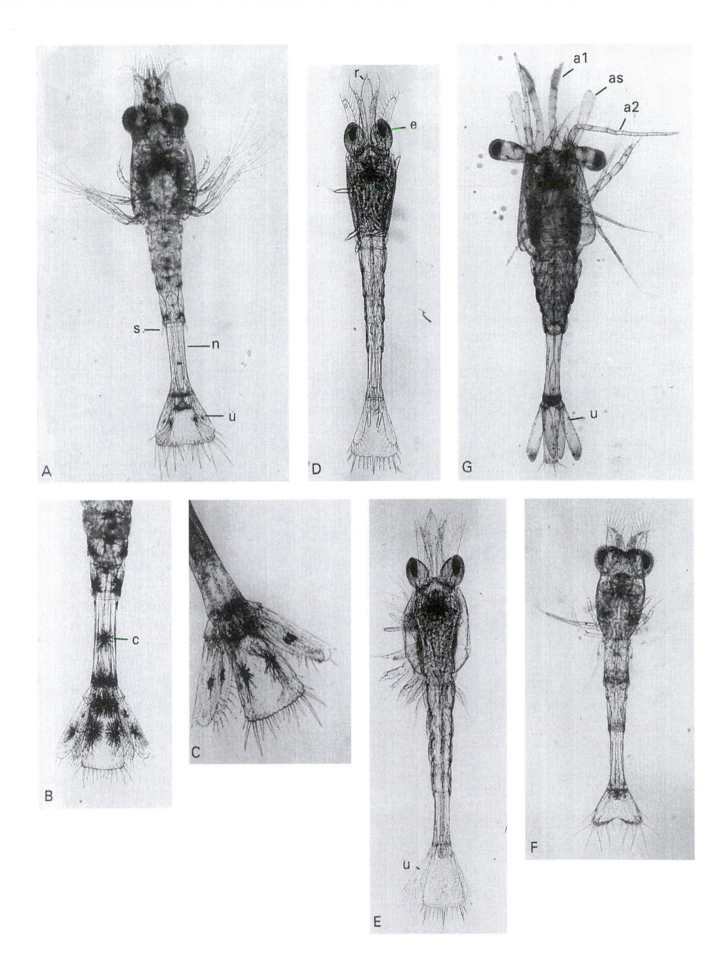

FIGURE 29: A Phylum Arthropoda, Class Crustacea, Subclass Malacostraca – Order Euphausiacea; B,D,E Phylum Arthropoda, Class Crustacea, Subclass Malacostraca – Order Decapoda (Section Caridea); C Phylum Arthropoda, Class Crustacea, Subclass Malacostraca – Order Decapoda (Section Macrura)

This figure illustrates a range of malacostracan larvae, including euphausiid, caridean and macruran representatives. They range from comparatively early larval forms (e.g. Fig. 29C) to almost fully developed individuals (e.g. Fig. 29E).

FIG. 29A. Unidentified euphausiid larva. [length 2.5 mm]. This larva is unusual in having a very short rostral spine, which is barely visible, and large mobile eyes. The long slender telson and uropods are also quite distinctive. These features together indicate that this is a furcilia larva of a species of euphausiid (see Fig. 26A).

FIG. 29B. *Processa nouveli holthuisi* Al-Adhub & Williamson. [length 4 mm]. This is a stage 5 zoea, which has distinctive antennules (a1), antennae (a2) and antennal scales (as), an elongate sixth abdominal segment and characteristically shaped and spinous telson and uropods. Note the elongated, stalked eyes (see also Fig. 28G).

FIG. 29C. *Upogebia* sp. [length 2.5 mm]. This is the first larval zoea stage, as indicated by the large, flattened, seemingly unstalked compound eyes and the poorly developed telson: no uropods are visible on the telson. The hindgut can be clearly seen, running the length of the abdomen (see also Fig. 24D).

FIG. 29D. *Philocheras fasciatus* (Risso). [length 3 mm]. This caridean larva is distinctive in having a rather darker (grey-brown) coloration than other similar crustacean larvae. The form of the telson, the outline of the abdomen (with its lateral spines), and the shape and disposition of the eyes and antennules are quite characteristic.

FIG. 29E. *Crangon crangon* L. [length 4 mm]. This is the fourth of five larval (zoea) stages of the common shrimp *C. crangon*. The thoracic limbs are already well developed, in contrast to the abdomen.

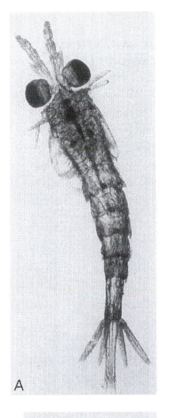

A

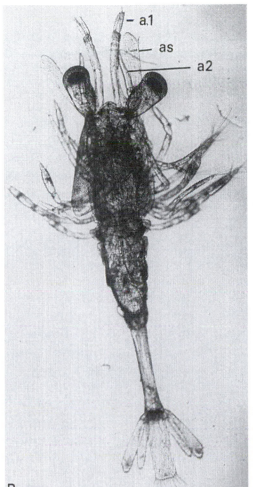

a.1

as

a2

B

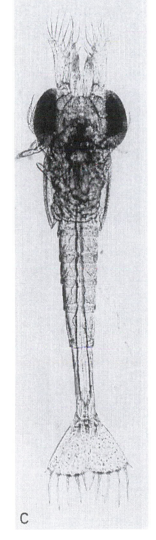

C

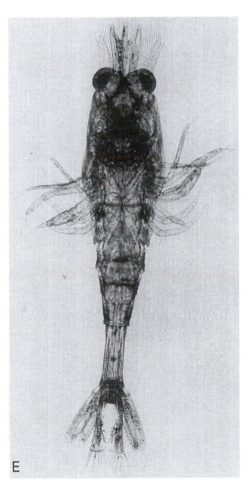

E

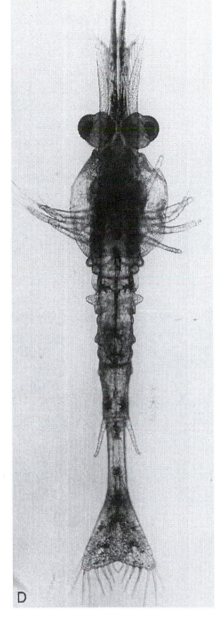

D

FIGURE 30: A–C Phylum Arthropoda, Class Crustacea, Subclass Malacostraca – Order Decapoda (Section Anomura)

This figure illustrates a range of larval stages of two species of anomuran hermit crabs. In general, anomuran larvae can be recognised (especially when viewed dorsally) by the more or less obvious pair of spines which project backwards from the lateral posterior margins of the carapace. Typically, amongst decapods, the adult female retains the eggs on the abdominal pleopods and the larvae hatch at a zoea stage; this lacks abdominal appendages and bears only ordinary thoracic (swimming) limbs.

FIG. 30A. *Pagurus bernhardus* (L.). [length 3.6 mm]. This is the stage 1 zoea. The characteristic feature of most Anomura zoeal stages is the presence of two posterior spines (ps) which project backwards from the carapace, and a more or less obvious (simple, anterior) rostral spine (rs). The form of the telson (t), and its pattern of spinulation, is usually species-diagnostic.

FIG. 30B. *Anapagurus hyndmanni* (Bell). [length 2.4 mm]. This also is the stage 1 zoea. Note the similar appearance of the larva, but the quite different telson in comparison to Fig. 30A (see also Fig. 33A,B; these show the fully formed juvenile of the related *Anapagurus laevis* (Bell)).

FIG. 30C,D. *Anapagurus hyndmanni* (Bell). [length 3.5 mm]. This is the stage 3 zoea. Note, in contrast to the telson (t), the considerable development of the thorax and the thoracic appendages (including the chelipeds, (ch) and the uropods (u)).

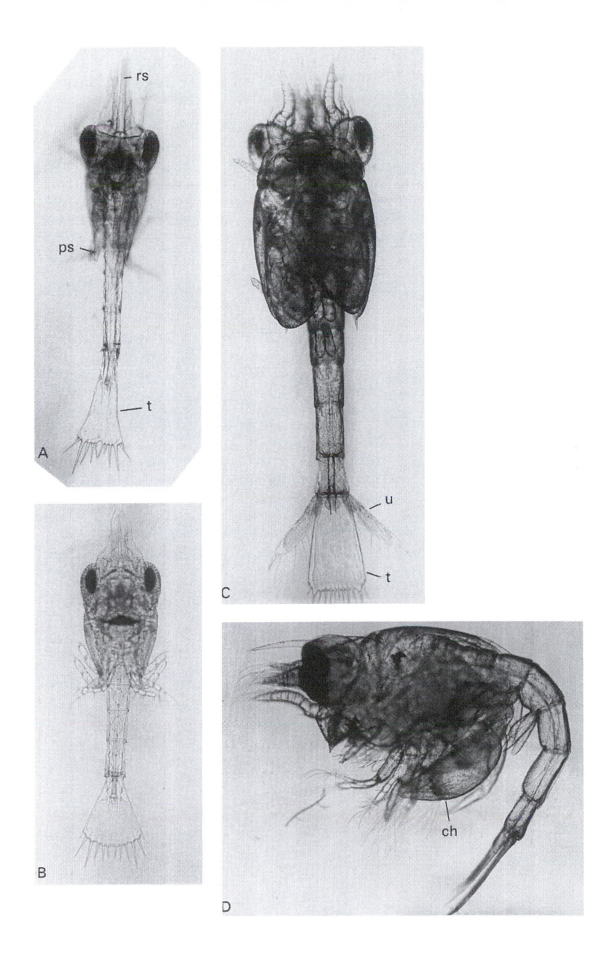

FIGURE 31: A–D Phylum Arthropoda, Class Crustacea, Subclass Malacostraca – Order Decapoda (Section Anomura)

The family Porcellanidae is represented by species of two genera in British waters, *Pisidia longicornis* (L.) and *Porcellana platycheles* (Pennant). Both have two zoea stages only and are very easily recognisable in plankton samples. The zoea bears a simple, very long, anterior carapace spine (sa) and shorter (but none the less exceptionally long) double posterior spines (dp) lying close together. In life the body appears to be suspended from the long rod formed by these spines.

FIG. 31A. *Pisidia longicornis* (L.). [length 10 mm]. This is the second zoeal stage, viewed dorsally. In this species the posterior spines are less than half the length of the anterior spine, often close to a ratio of 3:1. Characteristically there are bands of pigment on the rostrum and at the tips of the posterior spines. t, telson; ce, compound eye.

FIG. 31B,C. *Porcellana platycheles* (Pennant). [length 8 mm]. This is the zoea 1 in dorsal view. The double posterior carapace spines (dp) are at least half the length of the single anterior (sa) rostral spine. The anterior spine at its base has a distinct notch seen laterally. The distal region of the rostrum and of each posterior spine is pigmented. Note the setose thoracic swimming limbs (tl), and that the pleopods have not yet developed on the abdomen (ab). As the specific name suggests, the chelae of the adult are broad and markedly flattened, in contrast to those of *Pisidia longicornis*. t, telson.

FIG. 31D. *Pisidia longicornis* (L.). [carapace width 1.5 mm]. This is the (megalopa) post-larval stage. At this stage the larva may still be found in the plankton, with the pleopods providing power for swimming. The thoracic limbs (twl) now take on an ambulatory function. The antennae (a2) are particularly well-developed, in contrast to the antennules (a1), and the chelipeds (ch) are surprisingly large for an organism which is a deposit-feeder as an adult.

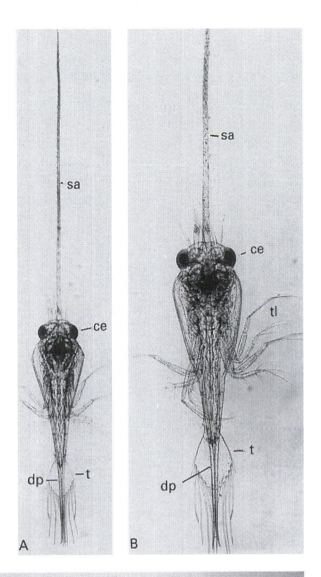

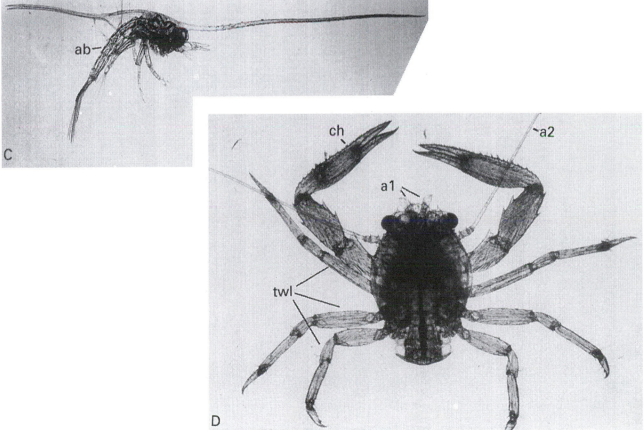

FIGURE 32: A,B Phylum Arthropoda, Class Crustacea, Subclass Malacostraca – Order Decapoda (Section Anomura)

FIG. 32A,B. *Galathea strigosa* (L.). [carapace width 2 mm]. These two individuals show the fully-developed juvenile squat lobster *Galathea*. These are free-living, symmetrically formed anomurans (cf. hermit crabs, page 62) which carry the abdomen reflexed beneath the cephalothorax as adults. The abdomen is, however, used in an effective tail-flip escape response, by means of which both the late larva and adult can rapidly swim backwards.

The chelae (ch) are long and setose, as are the three pairs of thoracic walking limbs (twl). The short fifth (l5) thoracic limb pair is held beneath the reflexed abdomen. The antennules (a1) and rostrum (r) are short, but the antennae (a2) are very long. The compound eyes (ce) are relatively large in the juvenile in comparison to those of the adult. The large tail-fan telson (t) and uropods (u) assist in the effectiveness of the tail-flip response.

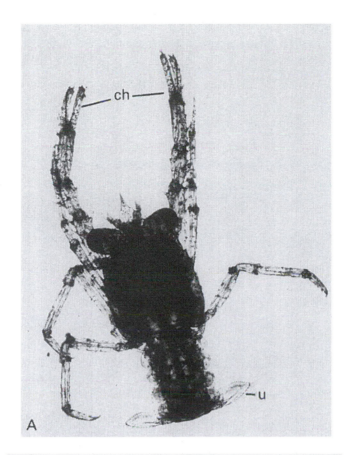

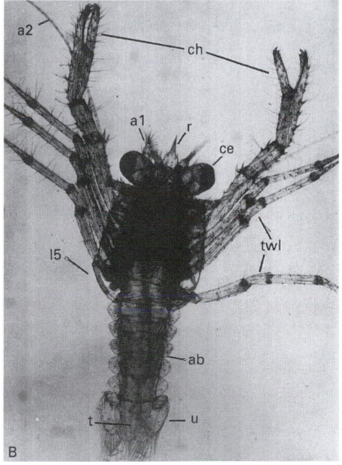

FIGURE 33: A,B Phylum Arthropoda, Class Crustacea, Subclass Malacostraca – Order Decapoda (Section Anomura); C Phylum Arthropoda, Class Crustacea, Subclass Malacostraca – Order Decapoda (Section Brachyura); D Phylum Arthropoda, Class Crustacea, Subclass Malacostraca – Order Decapoda (Section Macrura)

FIG. 33A. *Anapagurus laevis* (Bell). [carapace width 1.2 mm]. This is the fully developed juvenile hermit crab, ready to settle to the substratum. Note the asymmetry of the chelae (ch), the short antennules (a1), long antennae (a2) and the large symmetrical abdomen. The fifth thoracic limb (5tl) is also easily visible.

FIG. 33B. *Anapagurus laevis* (Bell). [carapace width 1.2 mm]. This is another individual, in which the left chela is folded beneath the thorax. Note the characteristic manner in which the fourth (4tl) and fifth (5tl) thoracic limbs project upwards when the abdomen is reflexed.

FIG. 33C. *Carcinus maenas* (L.). [carapace width 2.5 mm]. This is the megalopa of the common shore crab (see also Fig. 35C) shown here for comparison with the anomuran hermit crabs above (Fig. 33A,B). The symmetrical chelipeds (ch) and four pairs of thoracic walking limbs (cf. Fig. 33A,B) are completely developed, but the symmetrical abdomen (ab) has yet to flex beneath the cephalothorax. The antennae (a2) and large rostral spine (rs) between the stalked eyes are easily seen. The zoea of *C. maenas* is illustrated in Fig. 35A,B.

FIG. 33D. *Homarus gammarus* (L.). [length 9 mm]. This is the third zoeal stage. Lobster larvae are at an already advanced state of development at release of the first zoea: the female carries the eggs on her pleopods for several weeks before they hatch. The early larval stages of *Homarus* appear to be essentially pleustonic, but may also cling to floating debris, and are readily caught in pleuston nets. They seldom are caught in conventional plankton tows. Whereas most crustacean larvae are small-particle feeders, the zoeal stages of *Homarus* are carnivorous throughout. Following the fifth moult the juvenile status is attained and the individual settles to the substratum. Swimming of the later zoeal stages is primarily effected by the abdominal pleopods.

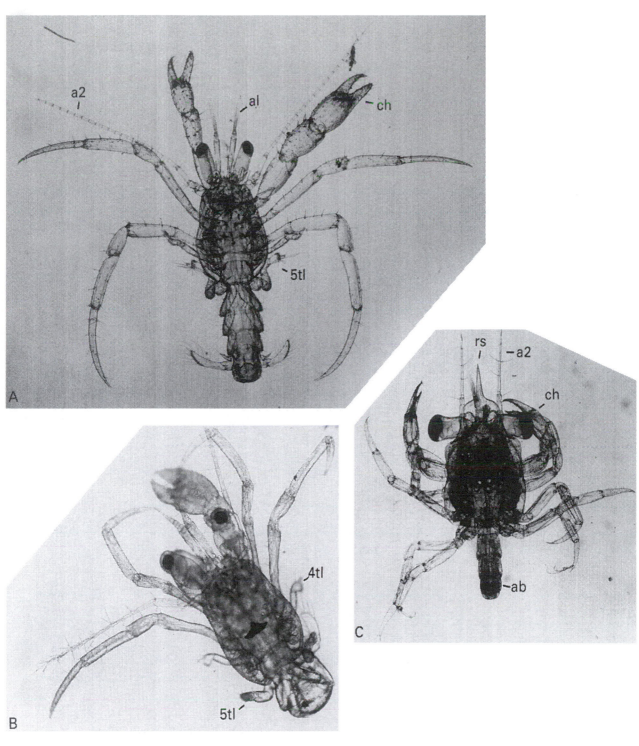

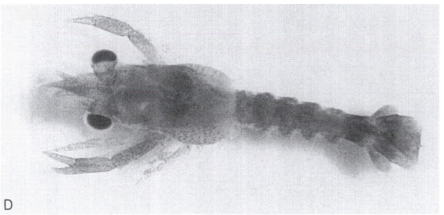

FIGURE 34: A,B Phylum Arthropoda, Class Crustacea, Subclass Malacostraca – Order Decapoda (Section Brachyura); C Phylum Arthropoda, Class Crustacea, Subclass Malacostraca – Order Cumacea

The decapod crustaceans generally release larvae to the plankton, albeit at varying stages of development. The higher decapods (e.g. Brachyura (crabs) and Macrura (lobsters)) release their young at a rather more advanced (zoea) stage, and have perhaps shorter planktonic phases of development, than do the more primitive malacostracans (e.g. penaeid prawns and sergestid shrimps). These latter freely spawn eggs to the water column and they hatch at a nauplius stage. Parental brooding of the embryos, to their advanced hatching stage, is the rule for higher decapods.

One curious feature of lobster reproduction is the infrequency with which their larvae are caught in plankton tows. By contrast, crab larvae – at the appropriate season – may be extremely abundant. Recently, however, it has been found that lobster larvae are virtually pleustonic, swimming very close to the sea surface and also perhaps clinging to drifting algae and similar objects.

FIG. 34A,B. *Necora puber* (L.). [length 3 mm]. This is zoea 3. This larva is quite large amongst brachyuran crabs and is recognisable from the long straight dorsal spine (ds) and rostral projections (rp), and the short lateral spines (lsp) of the carapace. The telson (t) is the typical fork-shape. The stalked compound eyes (ce) and elaborately setose thoracic limbs (tl) are evident.

FIG. 34C. *Diastylis rathkei* Krøyer. [length 3 mm]. Cumaceans are not truly planktonic, but despite being sand-burrowers they do swim quite effectively. Males are especially prevalent in plankton samples taken at night: females are, apparently less inclined to swim, but like the males they are periodically taken in deep plankton tows. The shape of the carapace, the swollen cephalothorax and elongate abdomen, and the characteristic form of the forked uropods (u) render cumaceans quite unmistakable.

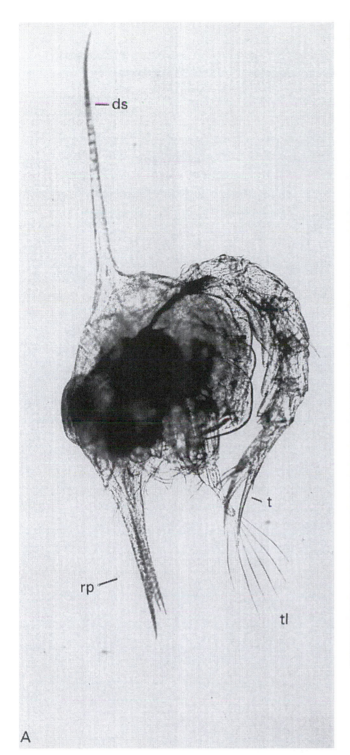

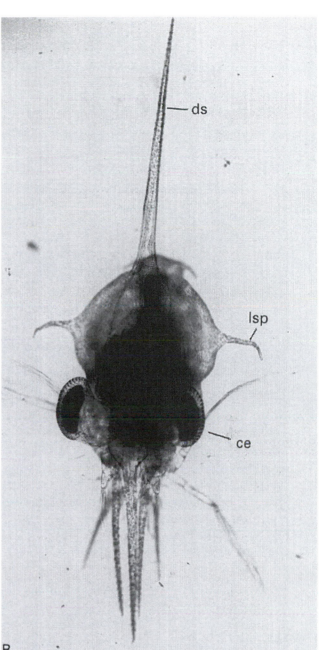

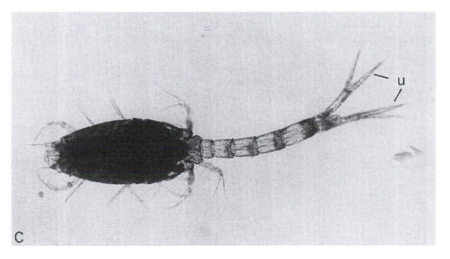

FIGURE 35: A–E Phylum Arthropoda, Class Crustacea, Subclass Malacostraca – Order Decapoda (Section Brachyura)

Brachyuran (crab) zoea larvae usually possess an anterior rostral spine and a single dorsal spine (but see, for example, *Ebalia tuberosa* (Pennant), Fig. 35D,E). There may be four or five zoeal moults/stages which finally give rise to the megalopa. The megalopa is the post-larval form and it differs markedly from the (laterally flattened) zoea in being dorso-ventrally flattened. On moulting to the megalopa stage brachyuran larvae also lose all trace of the dorsal spine.

FIG. 35A,B. *Carcinus maenas* (L.). [rostral–dorsal spine length 1.5 mm]. These photographs show the zoea 1 at different levels of focus. Fig. 35A shows the rostral projection (rp), whilst Fig. 35B shows most clearly the curved dorsal spine (ds). The fork-shaped telson (t), the abdomen (ab) (which does not yet bear pleopods) and the compound eyes (ce) are clearly seen.

FIG. 35C. *Carcinus maenas* (L.). [carapace width 2.5 mm]. This is the megalopa, in which the thoracic walking legs (twl), chelae (ch), compound eye (ce), antennules (a1) and antennae (a2) are fully developed. The rostral spine (r) is an obvious feature. The abdomen (ab) has yet to flex beneath the carapace, but otherwise this individual has more or less attained the adult form.

FIG. 35D,E. *Ebalia tuberosa* (Pennant). [length 1 mm]. This species has four larval stages; there are four zoea and the one (post-larval) megalopa stage. There are four pleopods (p) in the later zoea, and five in the megalopa. The antennules (a1) are present only as small projections. The zoea stages of *Ebalia* can be readily distinguished from other crabs by the virtual absence of a rostral spine and the smooth carapace, which has no dorsal spine. Moreover, the compound eyes (ce) are relatively large. Those illustrated are early zoea: Fig. 35E (zoea 1) shows no development of the pleopods, whereas Fig. 35D (zoea 2) shows only the early stages of development of the abdominal pleopods (p).

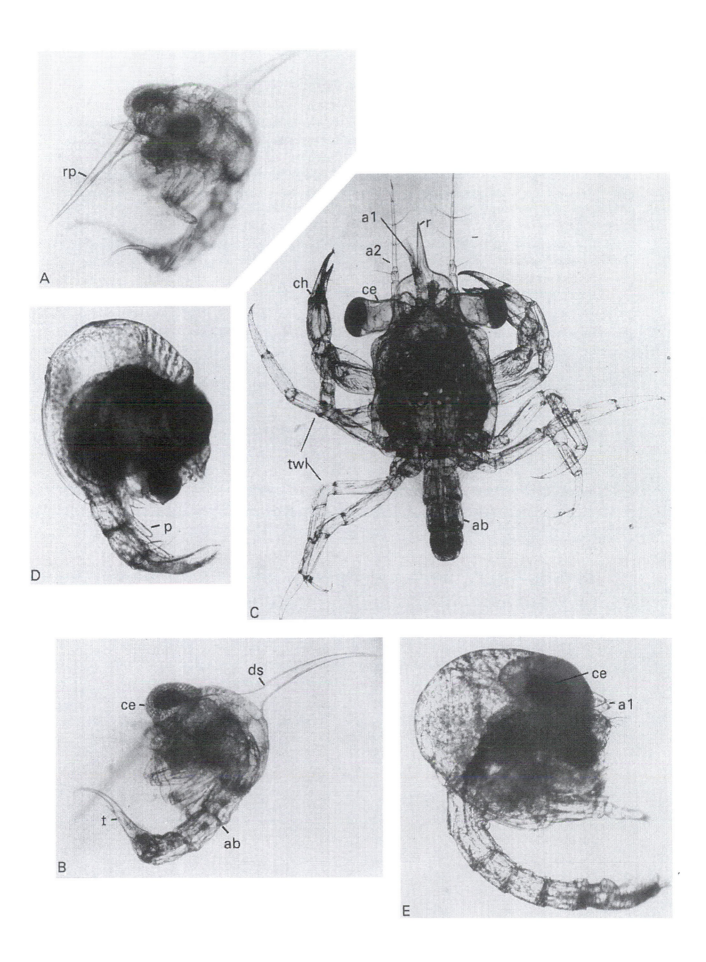

73

FIGURE 36: A–E Phylum Chaetognatha

The chaetognath 'arrow worms' are not related to the annelids, and derive their latinised name from 'spine jaws'. They are common, comparatively long-lived and widely distributed plankters, with various species found throughout the world's oceans. As such they have proved to be particularly useful as 'indicator' species of the origin or source of particular water masses. They are typically intermediate-sized predators and some species may attain several centimetres in length. In live samples they display exceedingly rapid forward darting movements interspersed with quiescent periods, during which they may drift downwards slowly. They prey largely upon small crustaceans such as copepods. Because of their almost total transparency they can also commonly be seen to be infested internally with surprisingly large nematode parasites.

FIG. 36A. *Sagitta elegans* Verrill. [length 23 mm]. This is a fully grown mature specimen, which clearly displays the hermaphrodite sexuality of this phylum. The ovaries (ov) project forward into the body cavity, anterior to the septum (s), while the testes (t) occupy the posterior region of the body: sperm can be seen to be extruded from the posterior lateral seminal vesicles (see also Fig. 36D). Around the British Isles, and subject to long-term fluctuations, this species is an indicator of offshore or oceanic water and is especially prevalent in the northern North Sea, and the Atlantic coasts of the west, but not necessarily the Irish Sea. In contrast to the appearance in this illustration, both live and preserved chaetognaths are usually straightened.

FIG. 36B. *Sagitta elegans* Verrill. [length 18 mm]. This specimen is not yet mature. The ovaries (ov) are very small, and the testes not evident. The anus (an) and septum (s) are, however, clearly visible and are adjacent to the lateral fin (lf).

FIG. 36C. *Sagitta elegans* Verrill. [length 22 mm]. The head of this adult specimen clearly shows the unusual external spinous (sp) jaws typical of chaetognaths. On engaging a prey organism the jaws fold inwards to force the item into the buccal cavity.

FIG. 36D. *Sagitta elegans* Verrill. [length 22 mm]. The small conical lateral seminal vesicles (sv) can be seen in their species-diagnostic position (cf. *S. setosa* Müller, Fig. 36E) – midway between the termination of the posterior lateral body fin and the caudal fin (cf), or tail.

FIG. 36E. *Sagitta setosa* Müller. [length 18 mm]. This species is generally rather smaller than *S. elegans*. Although the posterior lateral body fin (lf) and caudal fin (cf) of this particular specimen are somewhat disrupted, it can be seen that the seminal vesicles (sv) (which are asymmetrical in shape) are immediately adjacent to the termination of the lateral fin (cf. *S. elegans*). This is the most reliable means of discriminating the two species, although such a characteristic does preclude the identification of immature specimens. Somewhat unscientifically, these two species are, apparently, reliably separable on the basis of two further features. (1) *S. setosa* remains more transparent in formalin preservative than does *S. elegans*. (2) *S. elegans* remains stiff in preservative and is difficult to pick from a sample on a needle: *S. setosa*, by contrast, readily flexes across a needle. This species is characteristic of the central and southern North Sea and the eastern part of the English Channel.

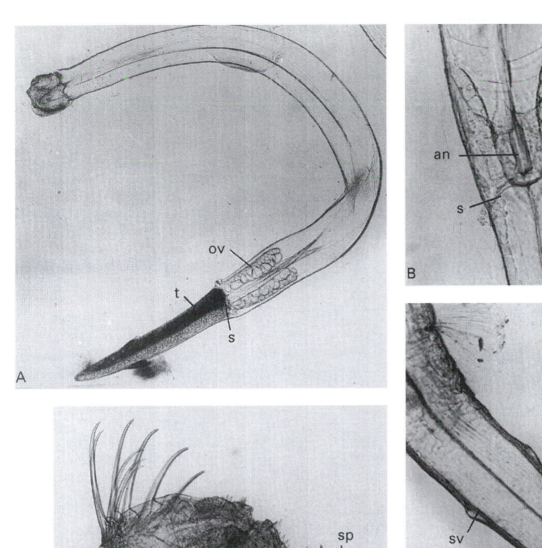

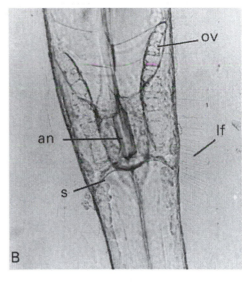

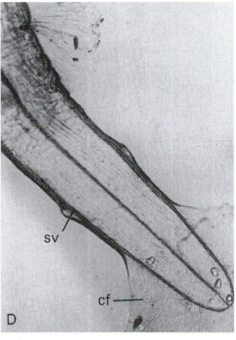

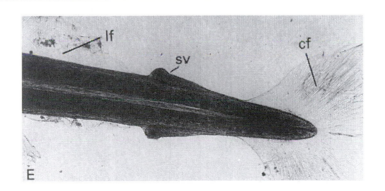

A,B Phylum Bryozoa, Class Gymnolaemata – Order Cheilostomata; C Phylum Bryozoa, Class Gymnolaemata – Order Ctenostomata

Bryozoan larvae are periodically abundant in plankton samples and the larvae of ctenostome bryozoan species may be quite common during the winter months. Metamorphosis of bryozoan larvae is generally rapid (a few hours) and gives rise to a zooid (ancestrula) from which other zooids bud to form the typical encrusting or arborescent bryozoan colony. The shelled cyphonautes larva (illustrated opposite) develops pelagically from eggs released by the adult polypide. This is in contrast to the shell-less, ciliated coronate larvae – more typical of the cheilostome bryozoans – which arise from internal fertilisation and which are brooded extra-coelomically in individual ovicells of the colony skeleton. It is curious that most observers would suggest that the cyphonautes larval form is 'typical' for the phylum, and yet it is displayed by only five genera worldwide. Of these, the genera *Membranipora, Electra* and *Conopeum* are all common around the British Isles. Amongst the cheilostome bryozoans the (shelled) cyphonautes larval forms are planktotrophic and are pelagic for perhaps three to four weeks: the coronate larval forms are, however, lecithotrophic and planktonic for only a few hours or days.

FIG. 37A,B. *Membranipora membranacea* (L.). [height 600 μm]. These are two examples of the cyphonautes larva of the common anascan cheilostome bryozoan *M.membranacea*. This larval form is typical of the calcareous gymnolaemates, in being laterally flattened and essentially triangular in shape. The soft tissues of the larva are contained between two hinged chitinous shell plates (sh), the notched apex of which comprises the sensory apical organ (ao). Swimming is effected by the ciliated bands (cb) of the two lobes. The larval gut (g) is particularly clearly shown in Fig. 37A. The pyriform organ (po), by means of which the larva attaches to the substratum at settlement, can be seen in both larvae. Metamorphosis into the ancestrula of the benthic adult colony is complete within a few hours of settlement. *Membranipora* larvae settle preferentially upon the macroalgal kelp, *Laminaria digitata* (Huds.) Lamour. To obtain a comparative impression of the size of this cyphonautes, see Fig. 46F.

FIG. 37C. *Flustrellidra hispida* (Fabricius). [length 900 μm]. The gelatinous (non-calcified) gymnolaemate bryozoans typically have unshelled larvae. The larva of *Flustrellidra* (Family Flustrellidridae) is, however, unusual not only in being shelled (reminiscent of the calcareous gymnolaemates), but also in being considerably more elongate and less flattened than its cheilostome counterparts (e.g. Fig. 37A,B). This larva is only briefly planktonic, but is, nevertheless, quite a powerful swimmer. The larvae settle preferentially upon fucoid macroalgae, especially *Fucus serratus* L., and metamorphosis to the ancestrula is rapid.

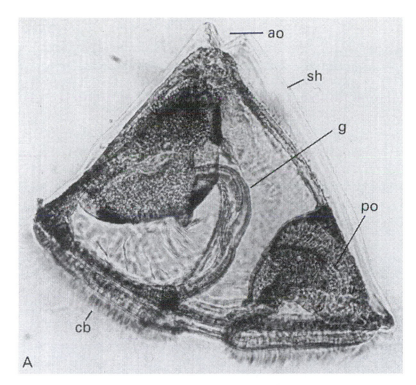

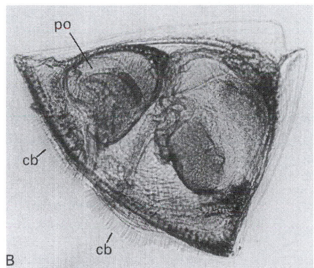

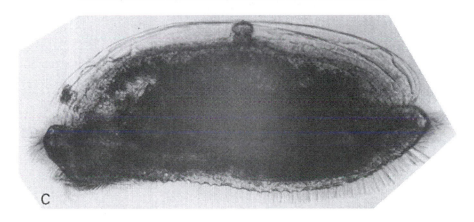

FIGURE 38: A,B Phylum Phoronida; C,D Phylum Chordata, Subphylum Euchordata, Class Osteichthyes

FIG. 38A. *Phoronis mülleri* De Selys Longchamps. [length 2.5 mm]. The phoronid worms are typically benthic in their ecology but the larval stage (actinotrocha) is planktonic, and may persist in samples for several weeks during the summer. Actinotrochs have a very spectacular and characteristic form with a pre-oral lobe (pl), or 'hood', extended into heavily ciliated fringing tentacles (lt). These larval tentacles are not, however, homologous with those of the adult (which develop beneath the larval structures ultimately to comprise the adult lophophore). The gut traverses the length of the larva and the anus (a) is terminal. Pigment spots – especially those associated with the glandular diverticulum (gd) of the gut – are obvious features of live actinotrochs.

FIG. 38B. *Phoronis mülleri* De Selys Longchamps. [length 3 mm]. The late larval stages of phoronids are remarkable for the type of metamorphosis displayed – the phenomenon of podaxonic development. Many invertebrate metamorphic changes are dramatic and that of *Phoronis* is no exception. The left side of the body develops a bud that elongates markedly: the developing adult gut (ag) (note: not the larval gut (lg)) loop is drawn out into this bud, but the anus remains in its anterior position. Finally the larval body is withdrawn and remoulded as the pre-oral lobe diminishes to form the epistome. The fringing epistomial larval tentacles and the terminal ciliated telotroch (t) vanish. The stage shown is about the latest that can be expected in plankton samples. The mouth (m), pigmented protonephridium (pn), terminal telotroch (t), larval gut (lg) and developing adult gut are all easily visible. In live specimens the gut contents can be seen to be rotated spirally as they are digested.

FIG. 38C. Larval fish. [length 3 mm]. Many pelagic and benthic fish spawn eggs that float, as a result of the high lipid content of the embryo. The yolk sac (ys) persists as a large ventral sac for some time, even when the fish hatches from the egg membrane. Only after having absorbed the yolk sac does the larval fish commence feeding.

FIG. 38D. Larval flatfish. [length 5 mm]. Flatfish (Order Pleuronectiformes), such as Plaice (*Pleuronectes*) and Dab (*Limanda*), undergo early larval development in much the same way as other fish. Asymmetrical growth and differentiation do, however, commence at an early stage in the larval life to give rise to the typical laterally twisted flatfish form. This can be seen in this figure by comparing Fig. 38C and 38D. Whereas the early larval stages are essentially transparent, the later larvae develop chromatophores (ch) in preparation for adult life.

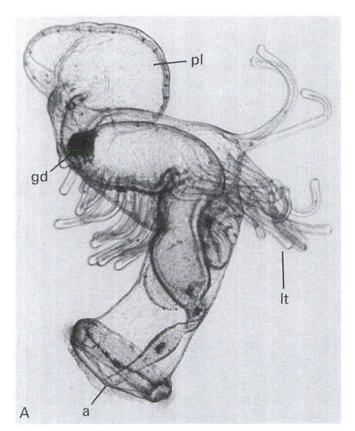

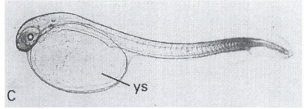

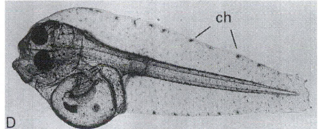

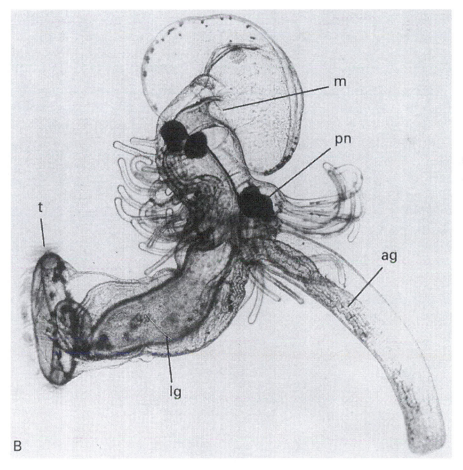

FIGURE 39: A,B Phylum Mollusca, Class Gastropoda, Subclass Opisthobranchia – Order Thecosomata; C Phylum Mollusca, Class Gastropoda, Subclass Prosobranchia – Order Mesogastropoda; D Phylum Mollusca, Class Gastropoda, Subclass Prosobranchia – Order Neogastropoda

This figure illustrates two species of mollusc which may be encountered in plankton samples. The first (*Limacina retroversa* Flemming, Fig. 39A,B) is holoplanktonic, being pelagic throughout its life cycle, and the second (*Lamellaria perspicua* (L.), Fig. 39C) is termed meroplanktonic because this species is planktonic only as a larva. The adult of *Lamellaria* is a benthic specialist predator of colonial ascidians.

FIG. 39A,B. *Limacina retroversa* (Flemming). [shell length 3 mm]. Fig. 39A shows the intact adult mollusc in the feeding mode, with the 'wings' (w) – lateral extensions of the foot – fully expanded. Swimming is effected by the rhythmical flapping of the wings, but these structures function also in feeding. The shelled, thecosomatous (cf. unshelled, gymnosomatous (Fig. 8A)) pteropod molluscs are particulate filter-feeders, trapping detritus and phytoplankters in mucus on the wings. Particles are drawn towards the wings by epidermal ciliary tracts on the wings and those entrapped in the mucus are moved by cilia to the central mouth. Reproduction is ovoviviparous and does not include a larval stage. The shell (Fig. 39B) is thin (light) and transparent and is tightly coiled. There is no operculum to close the shell following retraction of the body and the wings themselves cannot usually be completely withdrawn. Their major predator is perhaps the gymnosomatous pteropod *Clione limacina* Phipps (see Fig. 8A), for which *Limacina* is the major food item.

FIG. 39C. *Lamellaria perspicua* (L.). [shell length 0.9 mm]. This unusual mollusc larva is illustrated in its expanded appearance in Fig. 8C. The present individual is fully retracted into the shell, the unique double spire of which can be seen clearly. See Fig. 8C for further details.

FIG. 39D. *Nassarius reticulatus* (L.). [shell length 1.5 mm]. Although some of the more primitive molluscs possess an early larval form reminiscent of the annelid trochophore (hence indicating their evolutionary affinity), the typical larval form of the class Gastropoda is the shelled veliger (see also Fig. 8C,D). The larva of *Nassarius* has an especially well-developed velum (v), in which the characteristic two velar lobes are themselves elongated to give the structure a cruciform shape. The bands of marginal velar cilia (mvc) – by which the larva both swims and gathers micro-particulate food – are clearly seen. The whorls of the shell (sh) are seen in silhouette. Also, both the siphonal groove (sg) (confirming this as a species of the Order Neogastropoda) of the shell itself and the head tentacles (t) are visible. On being disturbed, the larva rapidly withdraws the velum and presumably sinks owing to the immediate cessation of active swimming. In this respect it should be noted, therefore, that veligers in preserved samples will *always* be retracted into the shell: all that can be recognised externally is the typical, albeit small, coiled gastropod shell.

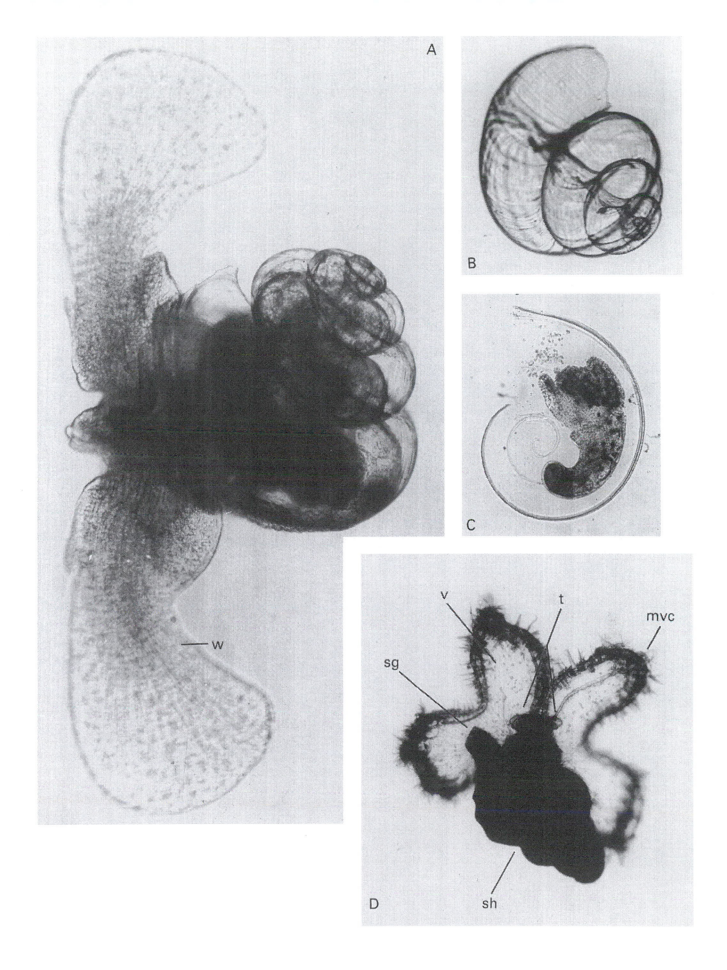

FIGURE 40: A Phylum Mollusca, Class Polyplacophora; B Phylum Mollusca, Class Bivalvia; C Phylum Echinodermata, Class Asteroidea

FIG. 40A. Unidentified chiton larva. [length 2 mm]. Chitons possess a briefly planktonic larva that is a rotund trochophore, of an appearance somewhat similar to that of a polynoid early nectochaete larva. Upon metamorphosis, however, the resultant juvenile chiton is elongate with the body thrown into folds. Anteriorly, the prototroch (p) persists for a while, sensory tentacles (st) appear, but the typical adult morphology is not yet discernible.

FIG. 40B. Unidentified bivalve post-larva. [shell length 1.5 mm]. Almost without exception the bivalve (or lamellibranch) molluscs reproduce by means of free-swimming shelled veliger larvae. Just as gastropod veliger shells appear similar to that of the adult snail, so also do those of bivalve larvae and adults. The velum of bivalve larvae similarly provides locomotory power as well as generating the larval filter-feeding currents. The bivalve larval shell is typically very thin and only lightly (if at all) calcified. It is transparent and the mantle cavity can be observed with ease. In this case the ctenidial (ct) filaments are visible, together with the foot (f), siphons (s) and visceral mass (vm). (see also Fig. 41E,F).

FIG. 40C. *Luidia ciliaris* (Philippi). [length 14 mm]. (see also Fig. 50D). This larva is much larger than that of any other British starfish and is large by any standard among marine invertebrates. It is probable that development of *Luidia* may require from several to many months before metamorphosis. At the stage shown here the adult rudiment (r) is beginning to form and the extensive ciliated bands (cb) are obvious.

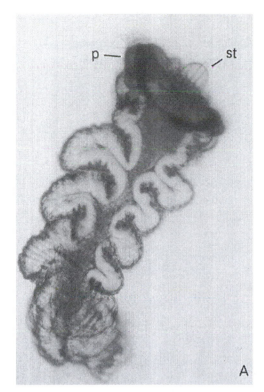

A

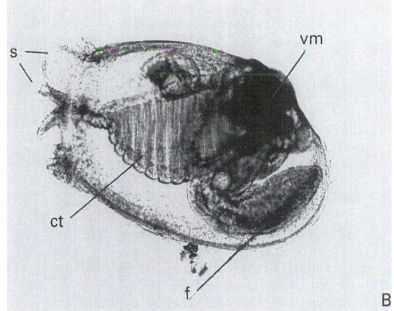

B

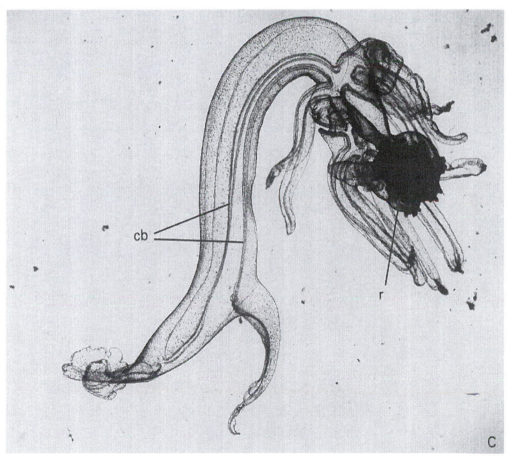

C

FIGURE 41: A,B Unidentified; C,D Phylum Mollusca, Class Gastropoda, Subclass Prosobranchia – Order Mesogastropoda; E,F Phylum Mollusca, Class Bivalvia

FIG. 41A,B. Unidentified trochophore larva. [length 100 μm]. This larva is probably of a mollusc, but it is possible that it may be attributable to a quite different phylum, such as the Sipunculida: several phyla display trochophore-like larval forms. This particular larva has a single large equatorial girdle of cilia (seen in plan view in Fig. 41B), and an unusually complex gut for so small a larva.

FIG. 41C. *Littorina littorea* (L.). [shell length 400 μm]. This is the late veliger of the common periwinkle. The ciliated velum (v) and head antennae (a), or tentacles, are easily visible. The fragile shell (sh) comprises the protoconch of the metamorphosed benthic juvenile.

FIG. 41D. *Littorina littorea* (L.). Egg capsule. [embryo diameter 150 μm]. *L. littorea* is unusual amongst molluscs in spawning individual egg capsules to the water column. Each contains at least one, but perhaps up to four, embryos (e). The bulk of the gelatinous capsule comprises a thin flat muco-polysaccharide disc (d), capped by a smaller hemispherical 'saucer' which contains the embryo(s). These float in the plankton and the embryos hatch as free-swimming veligers after six to seven days. Spawning of *L. littorea* appears to follow a marked lunar rhythm and individual adults may release eggs over periods of several months.

FIG. 41E. Unidentified bivalve juvenile. [length 500 μm]. This individual is fully developed and the ctenidium (ct) rudiments are clearly visible: that there is no trace of a velum shows that larval development is complete. The large foot (f) – with statocyst (st) – is already active, and the paired ctenidia are visible.

FIG. 41F. Unidentified bivalve veliger. [length 360 μm]. This is the typical appearance of inactive (or preserved) bivalve veliger larvae in plankton samples. The velum is contracted and the shell valves are closed.

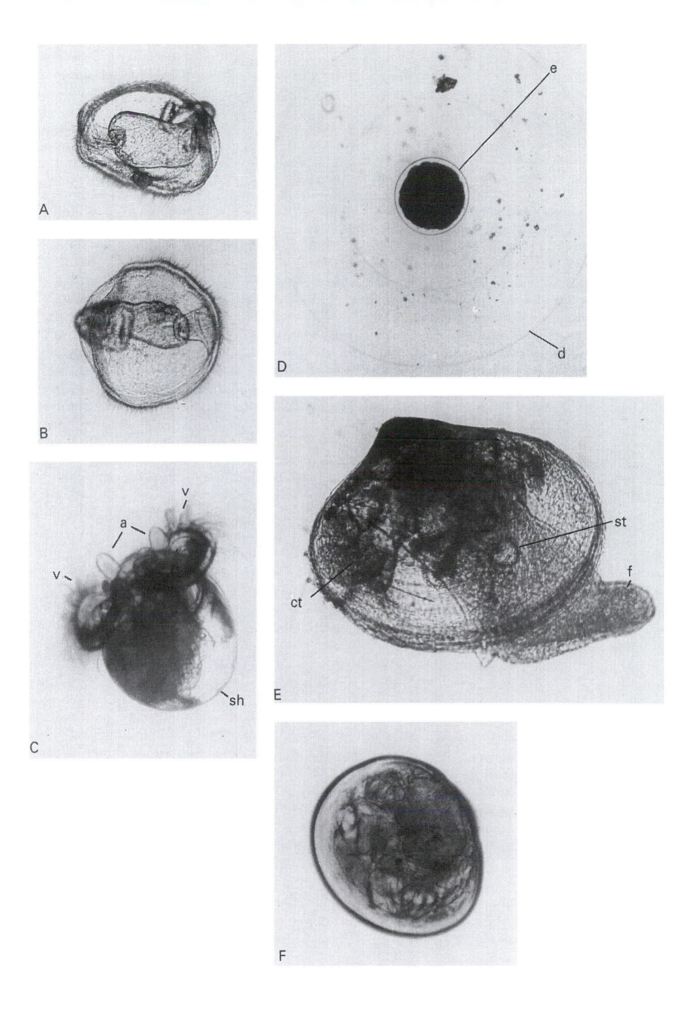

Cephalopods are often large very mobile predators. The benthic octopods range about on the substratum, leaving it to swim only briefly; cuttlefish spend much of their time buried just below the sand surface, but squids are almost continually active. Squids may be highly coloured, undergoing considerable and rapid changes by virtue of the many extremely active chromatophores lying in the epidermis. Such camouflage ability allows these cephalopods to match rapidly any background changes. Cephalopods lay large individual eggs, often bound together into characteristic arrays. The cuttlefish *Sepia* lays eggs which are blackened by the secretion of the ink sac and which are wrapped around algae and other protruding objects. Octopus, such as *Eledone,* lay groups of loosely associated eggs, whilst squids (e.g. *Loligo*) produce cigar-shaped spawn masses, each of about 100 eggs, in clusters attached to rocks and other benthic substrata.

Embryonic development takes a long time (five to six weeks) before hatching. Octopus possess a small tail fin at this stage and are planktonic only briefly before settling to the benthos for their adult life. Conversely, squid swim continuously from hatching. In view of their considerable locomotory powers cephalopod larvae and adults are only very seldom caught in nets.

FIG. 42. *Loligo forbesi* Steenstrup. [length 10 mm]. This figure illustrates a newly hatched squid with the yolk mass having been completely absorbed. Note the large eyes (e), short tentacles (t), huge brain (b), the mantle (m), ink sac (is), ctenidia (ct), visceral mass (vm), chromatophores (ch) and tail fin (tf).

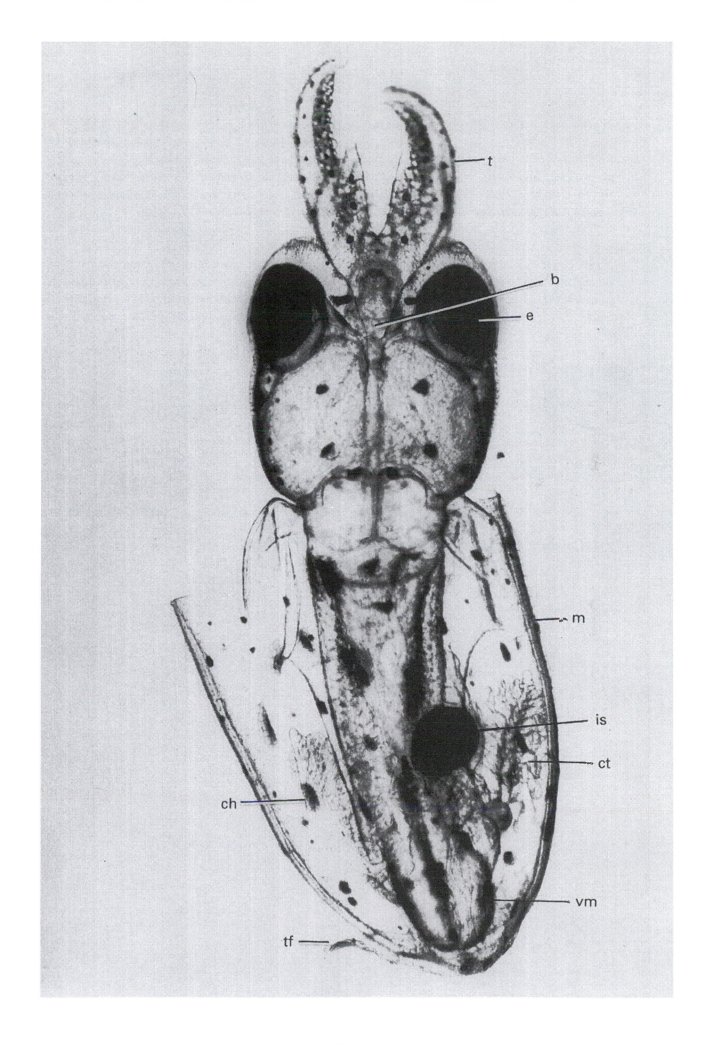

FIGURE 43: A–E Phylum Echinodermata, Class Asteroidea

Asteroid larvae are to be found in many stages in mid-summer plankton. Those shown here were raised in culture at a temperature of 15 °C and show the characteristic development stages of the common starfish *Asterias rubens* L. Other species have similar larvae and it is not always easy to distinguish them. The length of time taken to reach metamorphosis will depend to some extent on environmental sea temperatures, and hence the days quoted here may not be as occurs in nature.

FIG. 43A. *Asterias rubens* L., Day 6 larva. [length 250 μm]. This is the typical slab-sided early bipinnaria larva, with ciliated bands running around the periphery.

FIG. 43B. *Asterias rubens* L., Day 10 larva. [length 500 μm]. The bipinnaria has grown in length and width, the gut (g) is well defined and the primary coelomic sacs (cs) are appearing.

FIG. 43C. *Asterias rubens* L., Day 22. [length 1.5 mm]. The bipinnaria now takes on a more elaborate form, with the ciliated bands being thrown into longer projecting arms.

FIG. 43D. *Asterias rubens* L., Day 50. [length 2.5 mm]. This is the brachiolaria stage; metamorphosis will take place only when the rudiment has formed posteriorly and the larva has attached to the substratum. It will be seen that there are now three star-shaped, broad-ended, brachiolar arms (ba) which are the adhesive points by which the late larva attaches itself to the benthos.

FIG. 43E. *Asterias rubens* L., Day 60. [length 4.5 mm]. (see also Fig. 50C). The rudimentary fully formed juvenile (juv) is now in place, the brachiolar attachment points are fully developed and some of the longer arms are beginning to be resorbed. At this stage the larva is ready to adhere to an appropriate substratum and undergo metamorphosis. At metamorphosis the benthic juvenile breaks from the larval body.

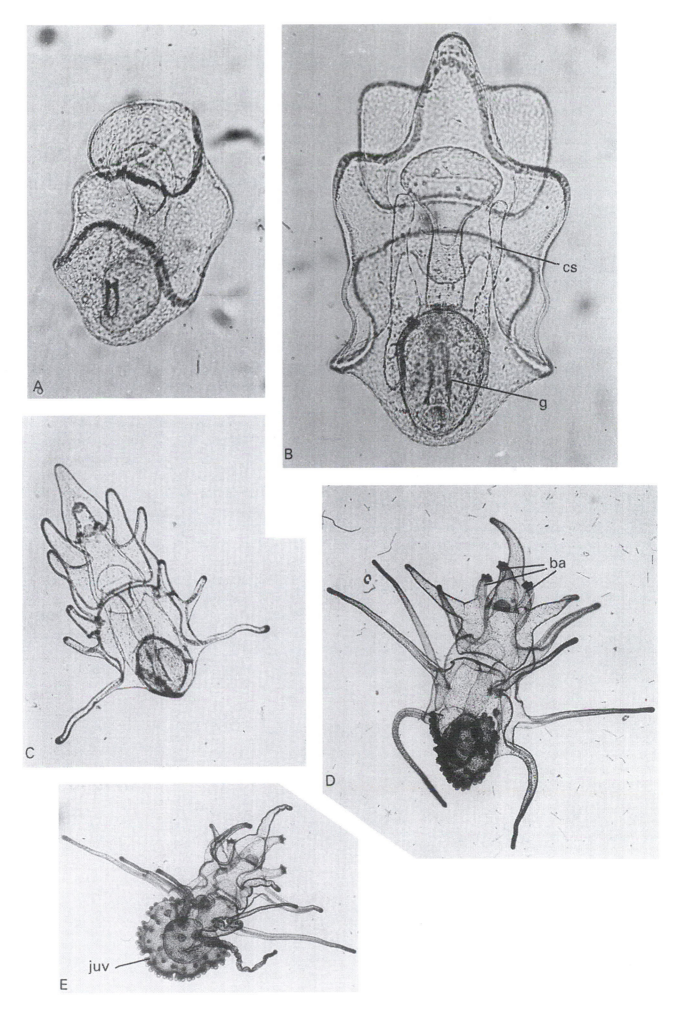

FIGURE 44: A–F Phylum Echinodermata, Class Ophiuroidea

The typical larval form of the classes Ophiuroidea and Echinoidea is the pluteus. In both classes this is characterised by more or less elongate 'arms', supported by skeletal spicules, bearing the ciliated bands by means of which both locomotion and feeding are effected. Ophioplutei are recognisable in being considerably more flattened and bilaterally symmetrical than echinoplutei (see Fig. 46).

It should be stressed that the larval arms are in no way homologous with the arms of the adult echinoderm: the latter develop progressively, as the tissues of the larval body are reconstructed, over a period of weeks. In all ophiuroids the juvenile disc eventually breaks away from the remaining larval body. This often occurs in the water column and it is not unusual to find fully formed juvenile brittle-stars in plankton samples.

FIG. 44A. *Ophiothrix fragilis* (Abildgaard). [length 3.5 mm tip-to-tip]. This is the fully formed pluteus just prior to the commencement of formation of the juvenile disc (see also Fig. 45B,C,E).

FIG. 44B. *Ophiothrix fragilis* (Abildgaard). [length 3.5 mm tip-to-tip]. These two individuals show the fully formed pluteus (above) and an individual in which the juvenile disc is almost completely developed (below). Note that the minor larval arms are no longer present in the more advanced individual: whereas the major larval arms remain intact, the minor arms undergo total regression as their larval tissues are broken down.

FIG. 44C,D. *Ophiothrix fragilis* (Abildgaard). [C, length 2.5 mm; D, length 2.1 mm tip-to-tip]. These are young, incompletely developed, plutei which show the details of the skeletal spicules and the pigment bands (pb), which are characteristic of *Ophiothrix fragilis*.

FIG. 44E. *Ophiura* sp. [length, 1 mm tip-to-tip]. In contrast to *Ophiothrix*, the pluteus of *Ophiura* is less elongate, more solid, and of a rather squat appearance. This larva clearly shows the form and articulation of the skeletal rods (sk), the central gut (g) region (around which the juvenile disc will form), and the terminal mouth (m) (see also Fig. 45D).

FIG. 44F. *Ophiura* sp. [length, 1 mm tip-to-tip]. The juvenile rudiment (rud) can be seen to be fully formed, because not only is the disc now complete, but also the adult arms can be seen clearly. Note that there is some regression of the minor larval arms (indicated by their asymmetry) but, as in *Ophiothrix*, the major larval arms remain intact.

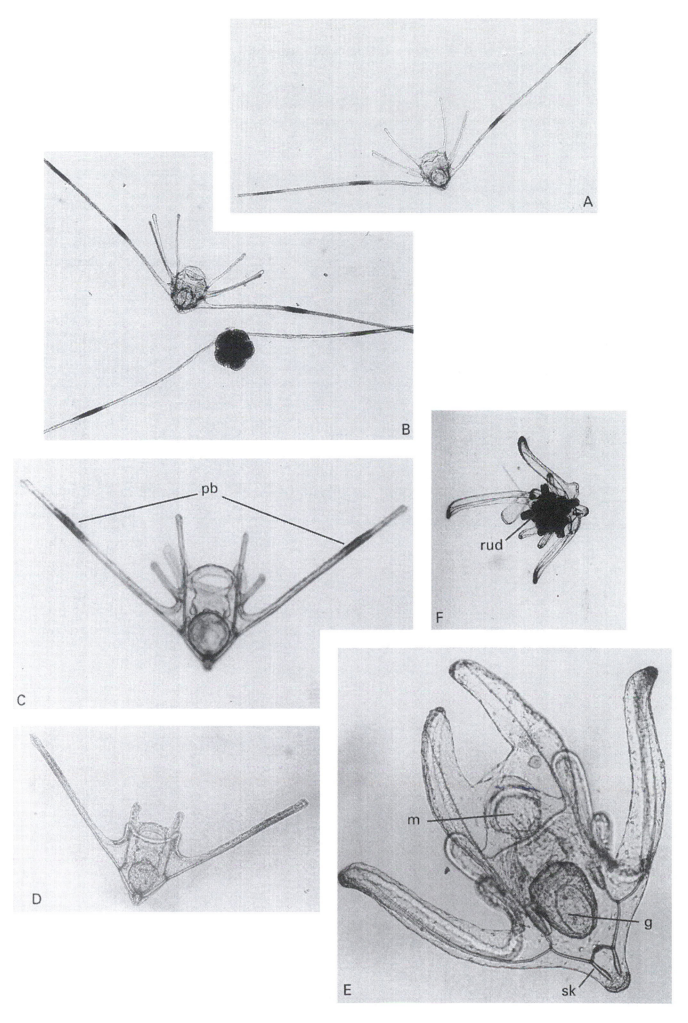

FIGURE 45: A–E Phylum Echinodermata, Class Ophiuroidea

These five figures illustrate the process of gradual development of the juvenile disc, and metamorphosis to the adult status, in three species of ophiuroid brittle-star.

FIG. 45A. *Amphiura filiformis* (O.F. Müller). [length 1.6 mm tip-to-tip]. As for *Ophiura* and *Ophiothrix* (in Fig. 44), the details of the larval skeleton are visible. Again, one can see that the two major larval arms remain intact throughout development of the juvenile disc. By contrast, all save one (broken) of the minor arms have regressed: the fragility of the minor arm in this instance indicates the resorption of skeletal materials.

FIG. 45B. *Ophiothrix fragilis* (Abildgaard). [disc diameter 350 μm]. Detail of the developing juvenile disc in an individual of similar developmental status to that shown in Fig. 44B.

FIG. 45C. *Ophiothrix fragilis* (Abildgaard). [disc diameter 400 μm]. Detail of the fully formed, detached juvenile disc.

FIG. 45D. *Ophiura* sp. This is the fully formed juvenile at the stage of becoming totally benthic in habit. Note the further development of the disc and juvenile arms, and the obvious tube-feet (t).

FIG. 45E. *Ophiothrix fragilis* (Abildgaard). [disc diameter 1 mm]. This is the fully developed juvenile of the common intertidal brittle-star. Note the more expansive and elaborate segments of the arms, in contrast to *Ophiura* (Fig. 45D).

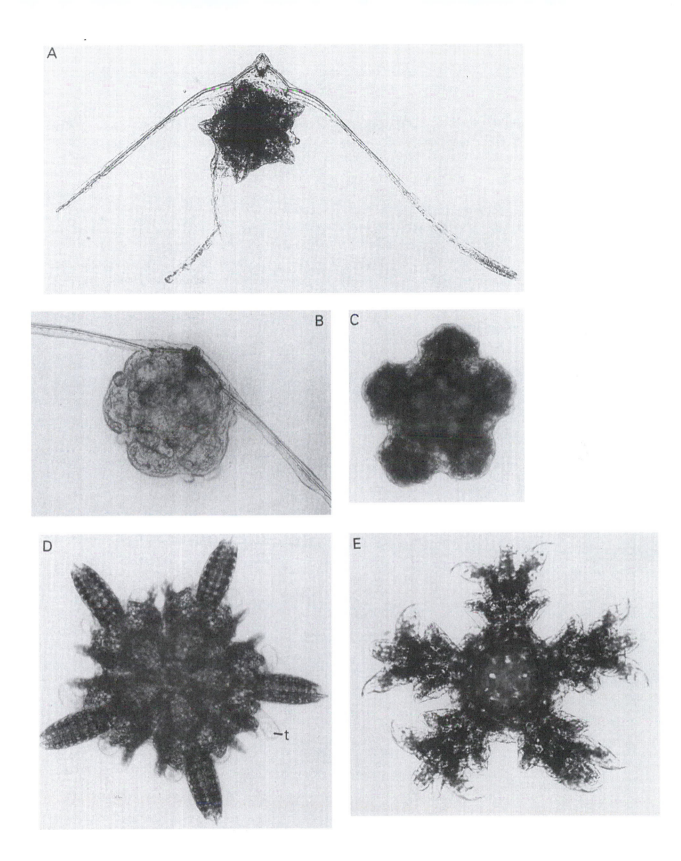

FIGURE 46: A–F Phylum Echinodermata, Class Echinoidea

This figure illustrates the pluteus larva of three common echinoid species. The echinopluteus differs from the ophiopluteus (see Figs. 44 and 45) in being more radially symmetrical, and usually bears a more or less obvious apical spine.

FIG. 46A,C,D. *Echinocardium cordatum* (Pennant). [A, lengths 0.4 and 0.6 mm; C, 1 mm; D, 1.5 mm]. These are examples of progressively older echinopluteus stages of the common sand-burrowing heart urchin *Echinocardium cordatum*. As for the ophiopluteus of brittle-stars, the juvenile urchin forms around the larval gut in a central position.

FIG. 46B. *Psammechinus miliaris* (Gmelin). [length 0.5 mm]. This is an early echinopluteus of this species – recognisable by the asymmetrical minor arms and the virtual absence of an apical spine.

FIG. 46E. *Echinus esculentus* (L.). [length 1.5 mm]. The echinopluteus of this species is unmistakable in having a swollen and spherical gut region, bearing equatorial lappets of ciliary bands (cb). This larva is quite transparent, by contrast to the rather heavily pigmented *Echinocardium* and *Psammechinus*. An unusual feature of this species' echinopluteus is the absence of an apical spine.

FIG. 46F. *Echinus esculentus* (L.). [diameter 0.5 mm]. This figure shows two fully formed juvenile urchins taken in plankton tows. The spherical test and the first spines are complete and the tube-feet (tf) are functional. Note the size of the adjacent cyphonautes larva of the bryozoan *Membranipora membranacea* (L.) [length 0.85 mm], by comparison (see Fig. 37A,B).

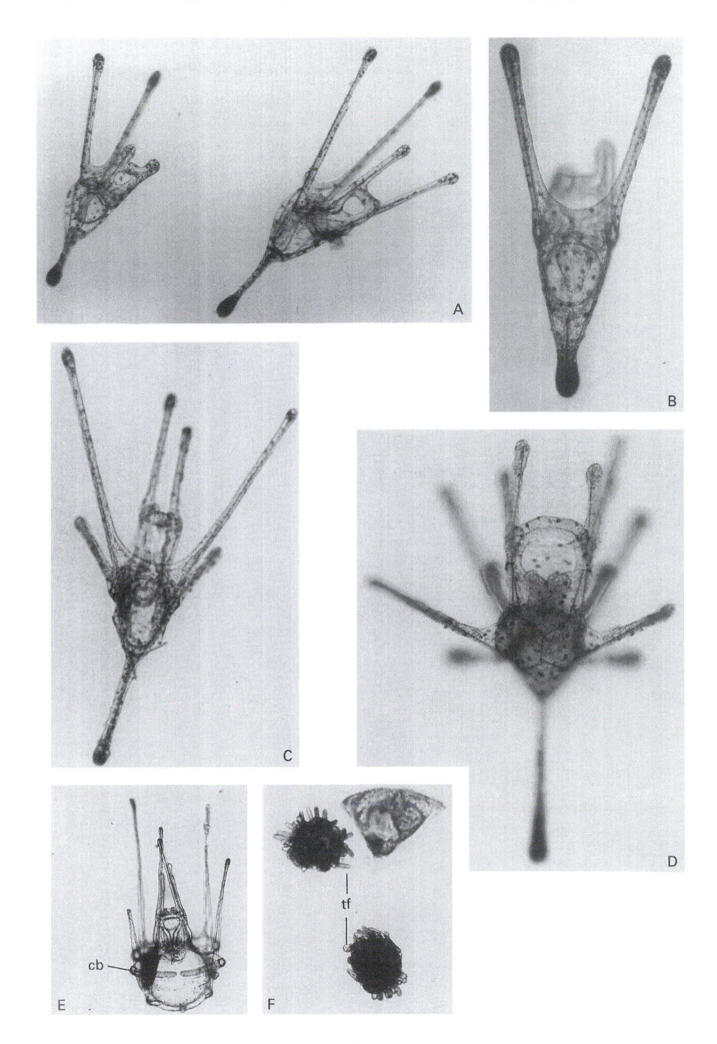

FIGURE 47: A–F Phylum Hemichordata, Class Enteropneusta

This figure (and Fig. 48) illustrates the tornaria larva, which characterises the enteropneust hemichordates. From its comparatively large size, transparency, gelatinous form, and the form of the ciliary locomotory bands, it is not difficult to see the affinity that these invertebrates are believed to hold with the Echinodermata. Tornaria larvae are lecithotrophic – that is, the larva is non-feeding and completes development utilising stored reserves provisioned in the egg – perhaps except for the oldest larval stage, which may feed. These larvae are not easily identifiable to species, largely because the adult has only seldom been ascribed to recognisable larval forms, and the tornaria invariably undergoes considerable, but gradual, change in the appearance and form of the ciliary bands as development proceeds.

Six hypothetical developmental stages are recognised and these are named after various zoologists. Stages 1 and 2 (the Müller and Heider) are barely distinguishable from the echinoderm auricularia, in being kidney-shaped and bearing simple ciliary girdles. Figs. 47 and 48 illustrate the four later stages (the Metschnikoff, Krohn, Spengel and Agassiz, respectively).

FIG. 47A–F. *Balanoglossus clavigerus* Delle Chiaje. [length 2 mm]. These tornaria larvae are almost certainly all various examples of the development of *B. clavigerus*, although it is possible that some examples of *Glossobalanus sarniensis* Stiasny are included. Fig. 47B is undoubtedly the Metschnikoff stage of *B. clavigerus*, viewed dorsally. The proboscis pore (pp) opening to the hydrocoel (h) – or proboscis coelom – can be seen clearly and the mouth is on the diametrically opposite (out of focus) surface of this larva. The spherical midgut (mg) and conical intestine (in) leading to the terminal anus (a) are evident in Figs. 47A–D, and the buccal cavity (bc) can be seen in Fig.

47D. In all cases the major propulsive ciliary girdle, the primary telotroch (pt) is obvious, as are the complex and convoluted ciliary bands (cb), which provide orientation.

The mesocoel cavity forms adjacent to the midgut at a late stage of development and the third coelomic cavity, the metacoel, forms around the intestine. Fig. 47F shows the final larval stage (the Agassiz), in which the proboscis is lengthening. Note that there is a distinct polarity of development: the proboscis is differentiated first, the 'collar' region of the hemichordate (incorporating the mesocoel) forms next followed finally by the 'trunk' region (incorporating the metacoel).

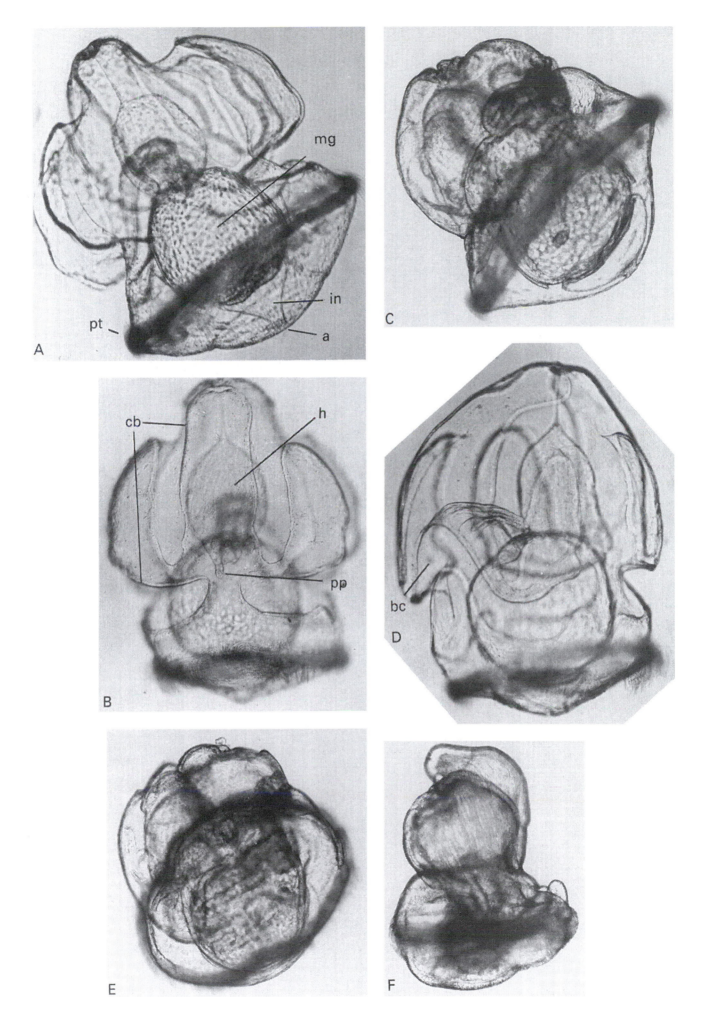

FIGURE 48: A,B Phylum Hemichordata, Class Enteropneusta

This figure illustrates two examples of the Metschnikoff stage of the tornaria larva of *Balanoglossus clavigerus* Delle Chiaje.

FIG. 48A. *Balanoglossus clavigerus* Delle Chiaje. [length 2 mm]. In this larva the hydrocoel (h), buccal cavity (bc), pharynx (p) and midgut (mg) are easily seen. Also the flashgun has 'frozen' the cilia of the primary telotroch (pt) such that the marked metachronism of the ciliary beat can be clearly seen.

FIG. 48B. *Balanoglossus clavigerus* Delle Chiaje. [length 2 mm]. This larva is at a slightly later stage of development than that in Fig. 48A. Note that the midgut (mg) and intestine (i) are relatively larger in size, and the anterior part of the larva is rather more swollen.

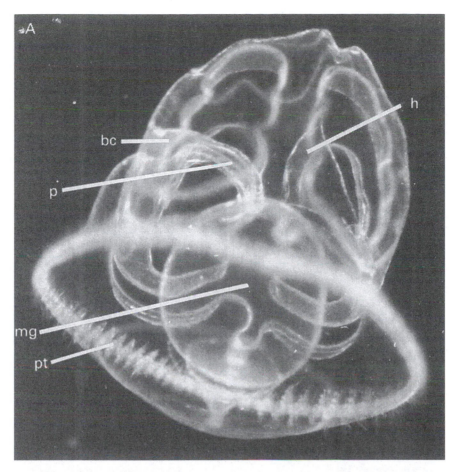

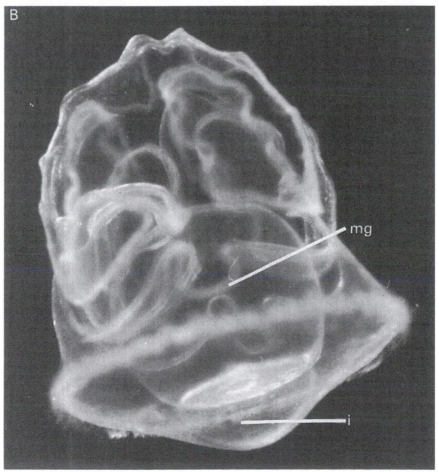

FIGURE 49: A–C Phylum Chordata, Subphylum Urochordata, Class Larvacea; D Phylum Chordata, Subphylum Urochordata, Class Ascidiacea

Two classes of the Subphylum Urochordata are entirely pelagic (i.e. 'holoplanktonic', cf. 'meroplanktonic' (see page 80)) throughout their life cycles. The first of these is the Class Thaliacea (which includes the Salps, see Fig. 50A,B) and the second is the Larvacea (or Appendicularia). As the name suggests, this latter class includes ascidians which retain larval features in the adult form (i.e. neoteny). All urochordates are filter-feeders, but individuals of the Larvacea are unusual in secreting a complex mucous 'house' about themselves. Within the 'house' the organism generates a feeding current by undulation of the tail, and edible particles are drawn in and trapped in a mucous net. The 'house' is extremely fragile, but easily re-built, and is always destroyed in netted plankton samples.

FIG. 49A. *Oikopleura dioica* Fol. [length of body, excluding tail, 1 mm]. This small larvacean is frequently seen in large numbers in plankton samples. Larvaceans are hermaphrodite, and in the present example, the well-developed ovary (ov) occupies the entire apical region of the body mass. The tail (t) is supported by the central core of vacuolated cells comprising the notochord (n). *Oikopleura* discards the mucous 'house' every few hours (possibly as a consequence of clogging of the filters). Sexual reproduction gives rise to the typical urochordate 'tadpole' larva (see Fig. 49D), but in larvaceans some of the larval features (especially the notochord and tail) are retained after metamorphosis.

FIG. 49B. *Oikopleura dioica* Fol. [length of body, excluding tail, 1 mm]. The individual eggs within the ovary can be more clearly seen in this illustration, as can also the endostyle (en) within the pharynx (ph).

FIG. 49C. *Fritillaria borealis* Lohmann. [length of body, excluding tail, 1.2 mm]. This lar-vacean differs from *Oikopleura* not only in terms of body shape, but also in behaviour. In contrast to the former genus, *Fritillaria* hangs outside and beneath the 'house'.

FIG. 49D. Unidentified ascidian 'tadpole' larva. [length 1 mm]. This is almost certainly the larva of a sessile solitary ascidian (e.g. *Ascidia, Ascidiella, Molgula*): 'tadpole' larvae of colonial ascidians are generally larger and less streamlined than that illustrated. The three adhesive papillae (ap), by means of which the larva attaches to the substratum prior to metamorphosis, are clearly seen. The only well-developed sensory organ of the larva – the so-called ocellus (o)/statocyst – is also evident. Garstang's Hypothesis relating to the evolution of the vertebrates revolves around the development of the 'tadpole' larva; this is proposed to have remained pelagic (by undergoing neoteny) as an adaptation to exploit the food-rich plankton. It is curious, however, that all ascidian larvae are strictly lecithotrophic, and none shows any trace of a larval gut.

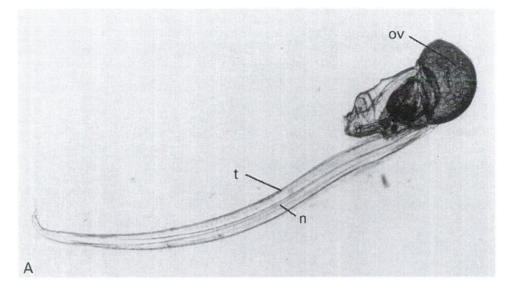

A

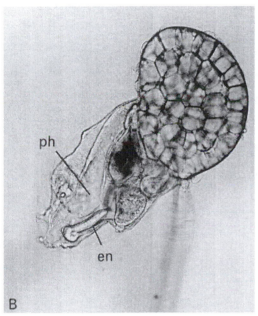

B

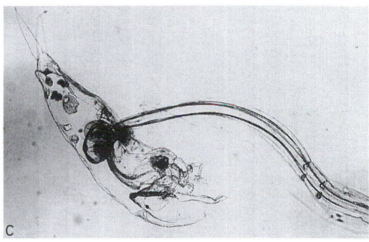

C

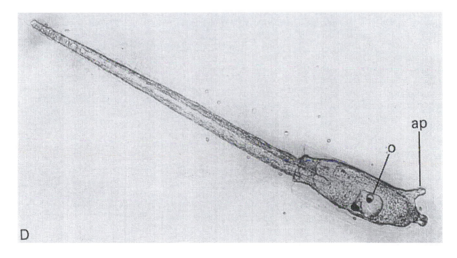

D

FIGURE 50: A,B Phylum Chordata, Subphylum Urochordata, Class Thaliacea – Order Salpida; C,D Phylum Echinodermata, Class Asteroidea

FIG. 50A. *Salpa fusiformis* Cuvier. [length 7 cm]. The pelagic Thaliacea are easily recognisable plankters. Individual *S. fusiformis* may reach 7 cm in length. By contrast, the sexual blasto-zooids, which are connected together, may form chains of zooids up to 1 m in length. In salp life cycles there is an alternation of gener-ation, with the asexual oozoids (as illustrated here) being solitary, having a stolon but no gonads. The branchial (b) and atrial (a) aper-tures are terminal anteriorly and posteriorly. The musculature is easily seen through the transparent test (t); the muscle bands (m) are characteristically incomplete ventrally. The endostyle (e) runs along the length of the pharynx which bears the filter-feeding gill-clefts. This particular species is an indicator of warm oceanic water, incursions of which into inshore areas may be heralded by vast swarms of salps. Oceanic samples of salp-dominated plankton frequently contain few other macro-scopic organisms.

FIG. 50B. *Salpa fusiformis* Cuvier. [colony 15 cm long]. (see also the solitary form illustrated in Fig. 50A). This is the colonial form of *S. fusiformis*, which may form enormous swarms comprising millions of organisms in which the chain-like colonies may attain up to 1 m in length. In this example there are only six zooids, each approximately 2.5 cm long, which are linked together by the stolon. Propulsion of the colony as a whole is by means of mus-cular contraction of the individual zooids.

FIG. 50C. *Asterias rubens* L. [length 4 mm]. This is the late brachiolaria larva of the common starfish *A. rubens*. Fig. 43 shows the develop-mental sequence of the larval stages of this species photographed with transmitted light. In this example the three brachiolar arms (by means of which the larva attaches to the sub-stratum, prior to separation of the fully formed juvenile; cf. Ophiuroidea) (ba) are clearly seen, and can be distinguished from the larval arms (la) by their terminal adhesive pads. The juvenile disc (jd) has almost com-pleted development and is beginning to close together, and the developing adult posterior coelomic pouches (pc) are evident.

FIG. 50D. *Luidia ciliaris* (Philippi). [length 13 mm]. With the exception of certain crus-taceans (such as stomatopods and spiny lob-sters), the larva of *Luidia* is perhaps among the largest of all marine invertebrates (occasional-ly attaining over 2 cm in length), and may require in excess of a year to complete devel-opment. It is spectacular not only in terms of size, but also by virtue of the extraordinarily complex larval arms and ciliary girdles. At the stage illustrated the juvenile rudiment (rud) is still small. As is typical for asteroids the fully developed brachiolaria larva settles on to the substratum, at which point the juvenile disc breaks away and begins benthic life. This sep-aration (rather than resorption) of the intact larval body raises the highly unusual possibili-ty of asexual production of genetically identi-cal juveniles from the one larva.

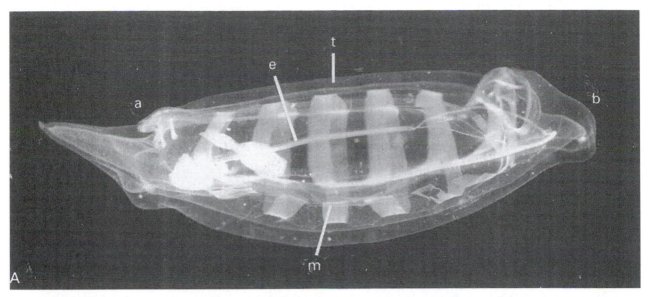

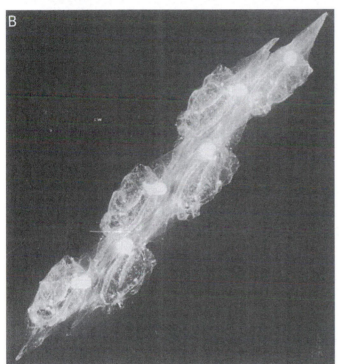

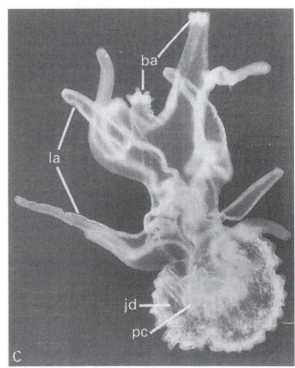

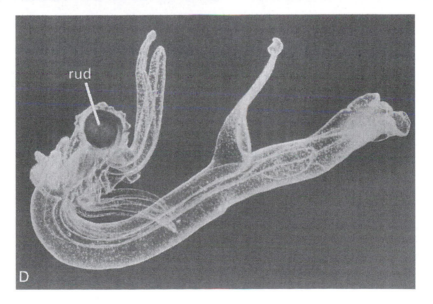

SPECIES INDEX

Organisms are identified by species. Headings given in **bold type** are not species but other categories.